ABSOLUT VODKA
Upper Temperature Threshold Exceeded
threshold
LASER MARK
lullaby
PAINTS
PREGNANCY & NEWBORN
h'eat it
PRESS HERE TO HEAT

Andrew H. Dent & Leslie Sherr

MCX
Material ConneXion®

MATERIAL INNOVATION
PACKAGING DESIGN

with over 350 illustrations

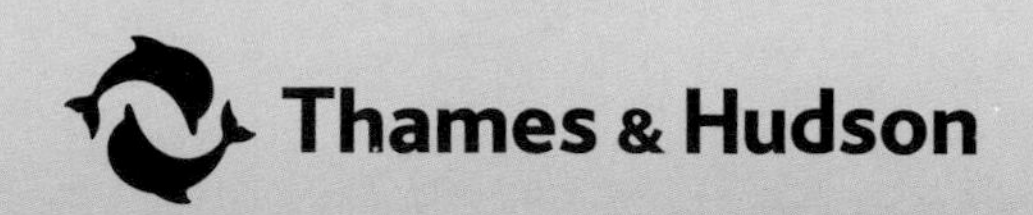

CONTENTS

For each chapter, a supplementary list of related materials has been included in the Materials Directory starting on page 180. The numbers are Material ConneXion's MC Index numbers which are a six-digit reference that is unique to each material in the library. The first four digits correspond to a specific manufacturer, in chronological order of acceptance into the archive. The last two digits are the chronological reference of each specific material from that manufacturer. The MC numbers are the basis for how searches are conducted through the thousands of materials that exist in the library. For more information visit www.materialconnexion.com.

First published in the United Kingdom
in 2015 by Thames & Hudson Ltd,
181A High Holborn, London WC1V 7QX

Material Innovation: Packaging Design

MCX
Material ConneXion
A SANDOW Company

Designed by Samuel Clark
www.bytheskydesign.com

British Library Cataloguing-in-Publication Data
A catalogue record for this book is available
from the British Library

ISBN 978-0-500-29197-9

Printed and bound in China by Shanghai Offset
Printing Products Limited

PREFACE

BY YVES BÉHAR, FOUNDER OF FUSEPROJECT

Packaging design is all the things we love, and loathe, at the same time. We love the discovery and surprise of unwrapping a new purchase, yet we hate seeing it so quickly discarded and turned to trash. The challenge of packaging design today is to accomplish the former without the hangover that comes with the latter. This is a schizophrenic task, but one that we, as designers, must reconcile in order to create an honorable and relevant new direction for our practice. When we succeed, we reveal design's capacity to change the course of an entire market category, raise people's expectations, show brands how to thrive by leading with creativity, and bring about a better future by inspiring other designers.

This does not happen in one go. It requires company leaders to demand compelling new sustainable and engaging solutions. They must support innovative ideas from internal and external creative teams, and empower manufacturing, sales, and logistics to push for the answers to those new problems. When we designed the Clever Little Bag for Puma, we spent three years understanding the world of global shipping, warehouse conveyor-belt systems, stockroom challenges, and the availability of different materials on the four continents where their factories are located. This took place while we were designing a new approach to the shoebox, with the goal of changing the customer experience. Our ambition was to make the box reusable with a low carbon footprint, while also improving the in-store stockroom experience. None of this was easy, nor obviously part of a designer's responsibility. But I see no greater accomplishment than when designers become partners to so many different stakeholders. In fact, without a partnership mentality, no real breakthrough can be achieved.

below The Clever Little Bag, designed by fuseproject for Puma.

opposite Design entrepreneur Yves Béhar, founder of fuseproject.

Over the next decade, packaging design needs to be reinvented. While it strives to realize the brand-building exercise it has been to date, it also needs to solve environmental expectations and new use cases. Some argue that most packaging design will "go away." I, however, predict that our involvement as designers will go much deeper toward material science, biology, engineering, and global logistics. This, combined with "people science" (how designers connect to consumers' emotional and physical needs), will lead to ever more extraordinary, smart, and magical packaging solutions.

What this volume dedicated to packaging—the third in an ongoing series focused on material innovation—does best is to underscore the designer's role in the acceleration and adoption of new ideas as we strive to reconcile our consumption with our environment. Packaging being the art form of presentation, we have the opportunity and responsibility to transform it into the very best that design can deliver.

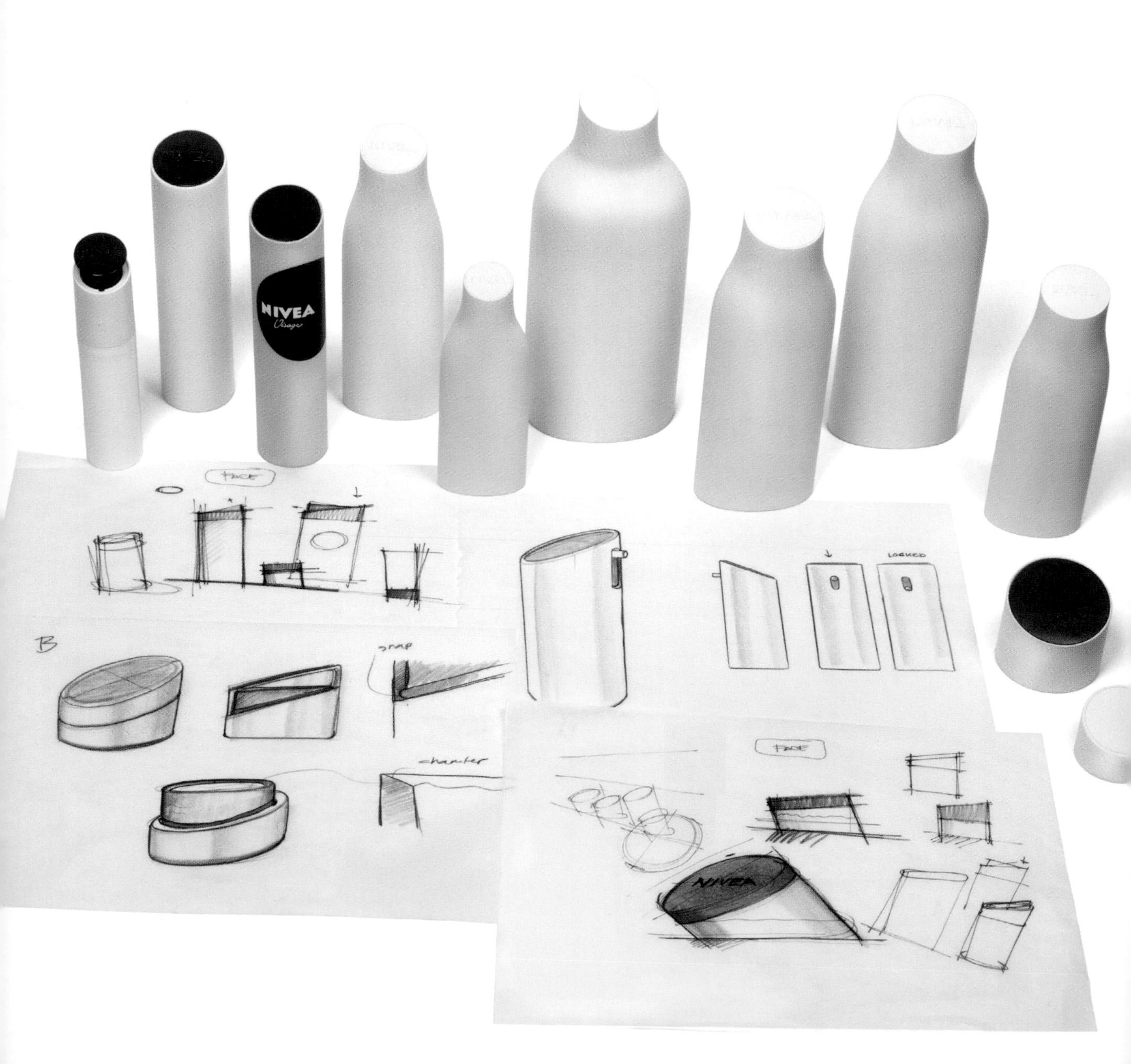
NIVEA
Visage
FACE
snap
LOCKED
FACE
NIVEA

INTRODUCTION

BY JOHN KIRKBY,
CREATIVE DIRECTOR, DESIGN FUTURES
SHEFFIELD HALLAM UNIVERSITY,
SHEFFIELD, UK

As the creative director of a successful structural packaging design studio, with a number of years' experience in commercial and packaging design research, I am in no doubt that successful contemporary packaging design relies heavily on good material choices. Both the functional and emotional aspects of packaging require careful consideration as to material specification. The packaging industry is broad and varied, and an immense range of materials and production processes is available to both producers and designers of modern-day packaging solutions and to the end users of these products, who are highly attuned to the visual and material languages of packaging with their attendant value systems. Consequently, designers need an in-depth awareness of available materials and an understanding of the implications of material choices, not just on packaging functionality and aesthetics but on the social and cultural perceptions around ideas of quality, freshness, environment and so on that come with these choices. Innovation can be achieved through clever implementation of existing materials, but nothing aids innovation better than the development of new ones. My design team are constantly on the lookout for the latest material developments. When a new material is identified, a process of explorative research involving the iterative design of packaging prototypes is undertaken to help gain an understanding of its scope and to push the creative boundaries of its use.

opposite For the overhaul of Nivea's primary packaging, fuseproject overlaid molded polymer bottles with visual and tactile references to the brand's iconic blue tin container and bold, Bauhaus typography.

Notwithstanding the ongoing need for innovation, packaging design for commercially successful products needs to perform the basics well. Fundamentally, good packaging needs to contain and protect the product while identifying or promoting the contents through either product visibility or graphic design. It goes without saying that this needs to be achieved at a viable cost. There are many considerations, however, that allow packaging to add value to a product.

For primary packaging in a retail environment, shelf stand-out is paramount. The emotional aspect of packaging plays a major part here. Graphics, photography, color, light, shape, and size are used to provide visual stimulus. Texture and shape can be used to encourage customers to reach out and pick up the product; even sound needs to be considered as an addition or at least a consideration in material choices. Crinkly plastic film is often regarded negatively, while the satisfying popping sound of a lid can often be seen as advantageous to the perception of a product's overall brand identity. Materials that speak to an environmentally aware agenda are also increasingly seen to hold consumer value. Creating a visual stimulus can prompt initial purchase, but design and material specification also have the potential to play an important role in promoting repeat purchases.

Once someone finds out I am a packaging designer, they rarely hold back from telling me what frustrates and annoys them about the packs they interact with. Stories about packaging that is difficult to open or reseal feature prominently, as do packs that create spillage or fail to contain their products. I find people tend to have strong, emotionally led views about packaging, and these views often guide their everyday purchasing choices. A well-constructed packaging-design process should start with the consideration of the end user of the product, and with some thought as to how they

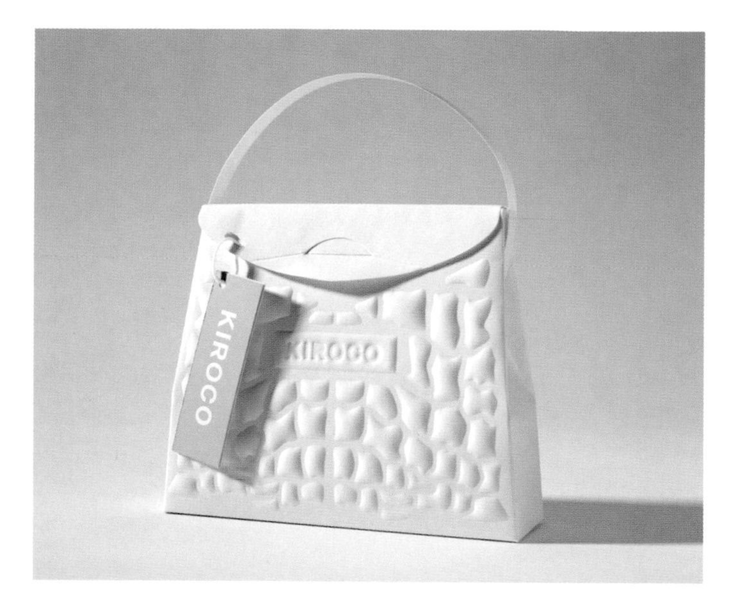

left The Paper Gift Bag for Kiroco, by Design Futures at Sheffield Hallam University, is made from BillerudKorsnas FibreForm material, a stretchable, formable packaging solution that allows for deep embossing to create unique shapes.

above An embossed detail of a FibreForm paper receipt holder, designed by Design Futures at Sheffield Hallam University, reveals the material's sculptural possibilities.

will interact with the pack. The process of designing a pack for pre-packed food to go—sandwiches, for example—might start by considering who the customers are and where they will be when they eat the sandwich. Also important is how they will open the pack, how they pick up the sandwich, what happens if they need to put it back down, and how they will dispose of the packaging at the end of its life. Considering this at an early stage of the design will help create packaging that a customer will appreciate and will purchase again. Shape, form, and materials are all important design tools to help achieve packaging that performs well with appropriate audiences and within particular contexts.

This "considered" design process is not important only for consumer products. Packaging designed to aid speed and efficiency in a manufacturing environment also needs to take into account human and process interaction. For example, innovative packaging solutions are used to supply component parts to production lines in the automotive industry, where efficiency of usage is key.

As mentioned earlier, when considering packaging-material choices, the environment also plays an important role. Companies of all sizes now recognize the value of packaging that addresses environmental issues. This approach may be led by pressure from legislation or by cost, but I find there is often a genuine desire by all the stakeholders in the packaging cycle to reduce the impact packaging has on natural resources and waste. Unquestionably, consumers are increasingly environmentally aware and will often avoid products that are over-packaged. Disposal issues also have an impact, and composite materials that are hard to separate for recycling or materials not widely recycled can affect purchasing decisions. There are some interesting material developments in this area as we begin to come to terms with the prospect of a limited supply of resources. Sustainable and responsible supply is also important, as is the development of materials that can replace existing non-sustainable packaging solutions. I have recently worked with a new paper product that is particularly strong, and research with this material has led to the possibility of considering paper where polymers were previously the only option. Once a paper-based option is explored it is surprising how many possibilities start to arise.

I believe it is a given that a designer should consider packaging material reduction and recycling options when specifying packaging. Reuse is also a very important consideration. Many packaging solutions have a useful life after their initial function. It is not always obvious what this will be at conception, but there is a particular skill in building reuse into packaging design from the start. Closed-loop packaging systems are also a way to promote reuse, and these usually require materials that are durable and able to withstand repeat use. Historical success in this area is the example of the returnable glass milk bottle; a local dairy owner told me recently that it was not unusual for a milk bottle to last twenty years or more through his traditional delivery and collection system.

Material developments are also of significant interest in adapting packaging design for a rapidly changing retail environment. Traditional shop-based retailing requires primary packaging to promote a product alongside its competitors, often in hugely congested environments. This means that packs have to work as hard as possible to sell the products inside them. The packs additionally need to protect and contain their products through the distribution and retail chain before being transported to end use by the customer.

As online retailing grows, however, the promotion function of the packaging starts to diminish. Products are increasingly selected through viewing online photography and video demonstrations, and the customer's first physical encounter with the packaging may not be until the product is delivered or collected—well after it has been purchased. This means that any visual presentation as well as the material choices are more a confirmation that the purchasing choice was a good one rather than an advertisement for the product. The protection and storage functions of packaging, however, may increase with this sales model, as products are often distributed through postal delivery systems that require different approaches to retail distribution. For example, pack orientation might change to fit through standard letterbox sizes, and graphics might move to the inside of a pack to provide an enhanced opening experience while concealing the identity of the product in the postage system.

As well as assisting in the areas I have already referred to, the future of packaging enhanced by new material development is an exciting one. Smart technology promises a future of packaging that is adaptive and customizable and can impart a whole wealth of information to the end user. This can be used to provide benefits in areas such as food wastage, process efficiency, and the administering of medicines. As long as designers keep up-to-date with ongoing developments for contemporary packaging in the scientific, technological, and social worlds, the opportunities and uses for new materials are exciting and endless.

above In a reinvention of clamshell packaging, the Jawbone UP24 fitness tracker sits recessed within a white, molded pulp form.

right A laser-cut paper "grille" interacts with the pattern on the inner sleeve of the Jawbone Era packaging, allowing the brand narrative to unfold along with the exterior packaging.

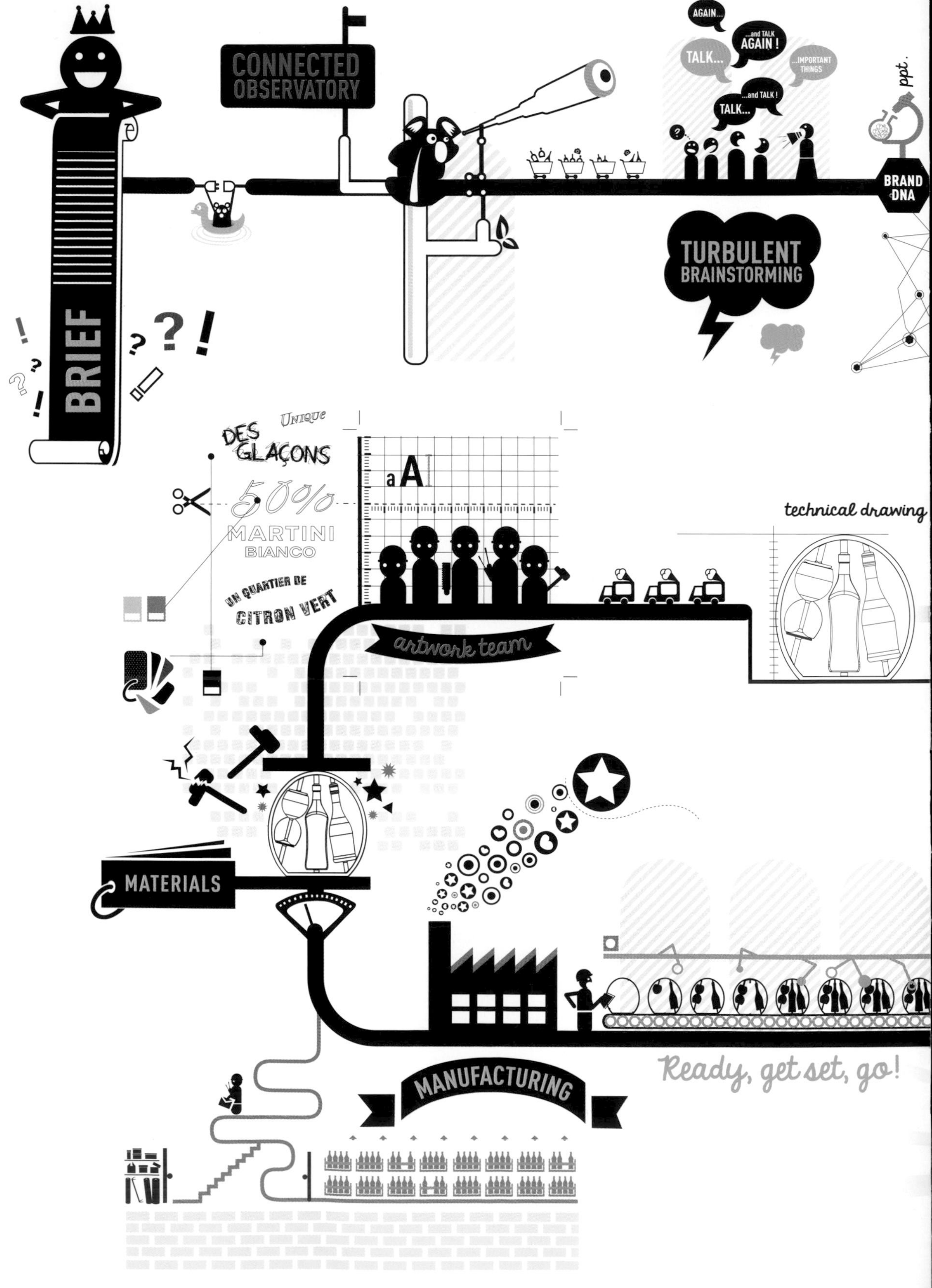

CONNECTED OBSERVATORY

AGAIN...
...and TALK AGAIN !
TALK...
...IMPORTANT THINGS
...and TALK !
TALK...

ppt.

BRAND DNA

TURBULENT BRAINSTORMING

BRIEF

DES GLAÇONS
UNIQUE
50%
MARTINI
BIANCO
UN QUARTIER DE CITRON VERT

aA

technical drawing

artwork team

MATERIALS

Ready, get set, go!

MANUFACTURING

WELCOME
to the Happy Creative Bar
logotype
typography
ai
pad
design
PRESENTATION
outside routes
clap
clap!
clap
CLAP
clap
idea
idea
idea
idea
TALK...
TALK...
AD
routes' room
1
2
3
Presentation round table
SELECTED IDEA
OPTIMIZATION
1
Final Mock-up
Buzz
the End
Shop
MARTINI Royale
MARTINI BIANCO
Distribution
LAUNCH
tasty!
Enjoy

CHAPTER 1
GETTING TO ZERO

Improved sustainability in packaging design has a simple challenge: What is the absolute minimum of materials and processing required to ensure that the product reaches its destination clean, fresh, undamaged, and with the correct information provided about its content? Anything beyond this is unnecessary, and places a heavier burden on the planet. The methods by which this can be achieved are many, as shown by the projects in this chapter; there is no single solution, but rather a consideration of each product's packaging needs based upon the different constraints it faces—"clean and fresh" has very different requirements for food compared with running sneakers. It is very much the case that "less is more" (to quote Mies van der Rohe) when considering sustainability in this category, whether that be in terms of material, shipping volume, ink for printing, or plastic use. Remember, though, that this can only be to the limit of good functionality; after all, returning a damaged product then reshipping a new one can hardly be considered low-impact.

Getting to Zero is really an examination of attempts to reduce packaging to its bare essentials, which in the best cases is literally nothing at all, or in some inventive cases is when the packaging itself can become the product (though these are the small minority). Safety, cleanliness, laws, customs requirements, and overall aesthetic appeal mean that most products will utilize some form of packaging material at some point in their journey from factory to foot or from farm to table, but the successful trend toward zero is a great thing for sustainability, and for good design in general. So how do we approach this goal? Nature of course

opposite Cocoform, a renewably sourced combination of coconut fiber (coir) and latex, is molded into complex forms to give flexible, durable cushioning, as exemplified by this egg carton.

leads the way, with eggs and bananas offering near-perfect case studies in shipping solutions (the egg does need a little help when going further than directly from coop to frying pan). The next generation of this natural solution is the laser-marking of fruit, which adds data directly onto the surface of the product using a fine laser etch, negating the need for those annoying little stickers. But what about a product that does not inherently come with its own packaging? Materials play a major role in this trend, aiding the creation of stronger, thinner, more easily biodegradable, and simply better-designed packaging as it heads ever closer to that ideal zero mark, but there are other ways to see how we can get to zero in our packaging needs.

An alternative definition of "zero" in this case might be no packaging waste, thus offering us solutions that reuse materials during transport, such as the returnable packaging used by furniture manufacturers to ship chairs and their parts, or reusable pallets, with strapping securing the products to them.

It can also mean zero glue, volatile organic compounds (VOCs), or virgin materials, the reduction of each of which is a move in the right direction. The method by which cardboard packaging is constructed to maintain form without the need for gluing has been taking its cue from mathematical solutions for origami. The much greater selection of certified "clean" recycled plastic streams for use in both primary and secondary packaging has meant that recycled content in a bottle or jar can now be specified with no compromise on cost or performance.

As with so much of the innovation in this sector, it is essential that the new materials be integrated at the concept stage, utilizing their unique properties to enhance and inform the design. Direct material replacement rarely ever leads to improved performance (or to zero anything) without the accompanying sympathetic design. Removing the packaging from the municipal waste and recycling streams through company-sponsored "Take Back" programs enables closer control and more effective recycling of some forms of primary packaging (when the company recycles its own product, it can more easily guarantee quality), though to date these have been only sporadically successful, with reasonable questions thrown up about the overall impact of such transport-heavy processes.

Clearly, the idea of "Getting to Zero" as a mantra for reducing the impact of packaging has a broader impact than simply thinner-walled water bottles with smaller caps. Ultimately, the most effective approach for creating solutions in this area is to consider packaging as a system rather than a container. Only when we understand the way in which the packaging works as a tool for shipping the product can real innovation take place. An instructive example of this is the

Packaging-like form is given strength and structure simply by the folds created in its surface. This piece was created using the ORI-REVO software designed by Professor Jun Mitani (see **opposite**).

This paper has been folded in accordance with schematics generated by Professor Jun Mitani's ORI-REVO software to create 3D structures through origami.

Clever Little Bag concept used to package Puma sneakers. Utilizing a bag that protects the products from the factory to the consumer's home, with cardboard used only from the factory to the retail outlet as a protective structure, reduces materials, cost, weight, and volume. This success was possible only once all the requirements of the packaging were understood, so that reductions could be addressed across the entire supply chain, ensuring that minimizing all unnecessary aspects of the package did not compromise its successful performance as a tool for getting the sneakers to the customer's home efficiently, safely, and beautifully.

Ultimately, the idea of removing all packaging is merely a ruse to get us to think about what is essential, and to discard all else. The journey is what matters, along with what it teaches us about our responsibility to good design and a less cluttered planet.

GLUELESS "ORIGAMI FOLD" CARTON

DESIGNER
Josh Ivy, Pangea Organics
www.pangeaorganics.com

DESIGNER BIOGRAPHY
As senior creative at Pangea Organics, an organic company based in Boulder, Colorado, Josh Ivy has used design to attract customers, earn accolades, and grow the business, all while protecting the Earth. His visual influence extends from overall brand image to new product design and direction, reinforcing the company's philosophy of sustainable and organic living. DuPont recognized the company with a Packaging Innovation award, as did the judges of competitions sponsored by *Print*, *How* and *Graphic Design: USA* magazines.

MANUFACTURER
All Packaging Company
www.allpack.com

MATERIAL
WindPower 80 Recycled Board
and vegetable-based ink

MATERIAL PROPERTIES
Recyclable, Recycled Content, Simplification, Secondary Packaging

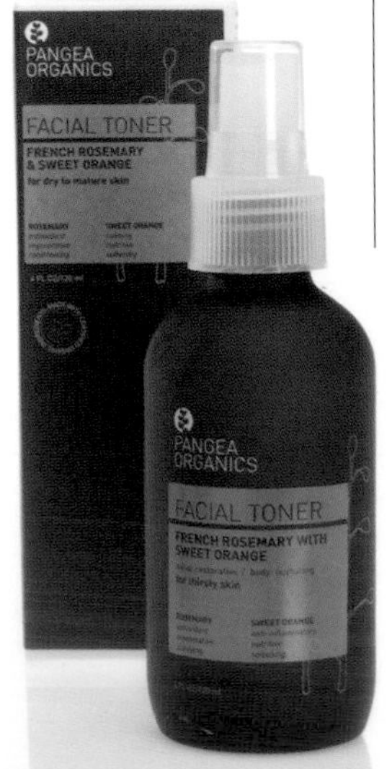

The rich orange-brown of amber glass, a color and a material preferred by producers of essential oils and pure plant ingredients for its ability to shield fragile contents from sunlight's UV rays, extends from bottle to carton.

There is little that is truly new about a glueless paper carton. The "origami fold" packaging created for Pangea Organics, an advanced line of organic skincare products, has its origins in sixth-century Japan, when the practice of paper-folding was transformed from the exploration of a rare material into an enduring art. From *obi* (sashes) to *furoshiki* (bundles) to *shoji* (screens), the Japanese are masters of layers and wraps, elevating humble objects through the sophisticated use of natural materials. From pleating to folding to crumpling, their traditional boxes, baskets, and wrappers showcase handwork for functional as well as ritual use.

Designers of all disciplines have been inspired by origami's precise forms and mastery of material. For Josh Ivy, senior designer, it was an approach perfectly suited to a 100 percent natural product whose packaging needed to convey a commitment to the use of post-consumer waste, and which would itself be 100 percent recyclable.

Intended to minimize the overall packaging, simplify the box assembly, and connect with consumers, the tab-lock design easily releases to reveal a chocolate-brown printed interior that coordinates with the light-protective bottle and provides a surface where product information is conveyed, as well as the merits of the brand's structural solution. Laid out flat, the carton's carefully conceived outline makes its own case for quality innovation in sustainability.

MATERIAL INSIGHT

In a way, this packaging shows how there is nothing necessarily innovative about a good, sustainable solution. Pangea Organics decided early on that it would source the lowest-impact materials and produce the simplest functional structure for what it needed. That this award-winning combination is still at the forefront of its field is a testament to the company's commitment to minimizing

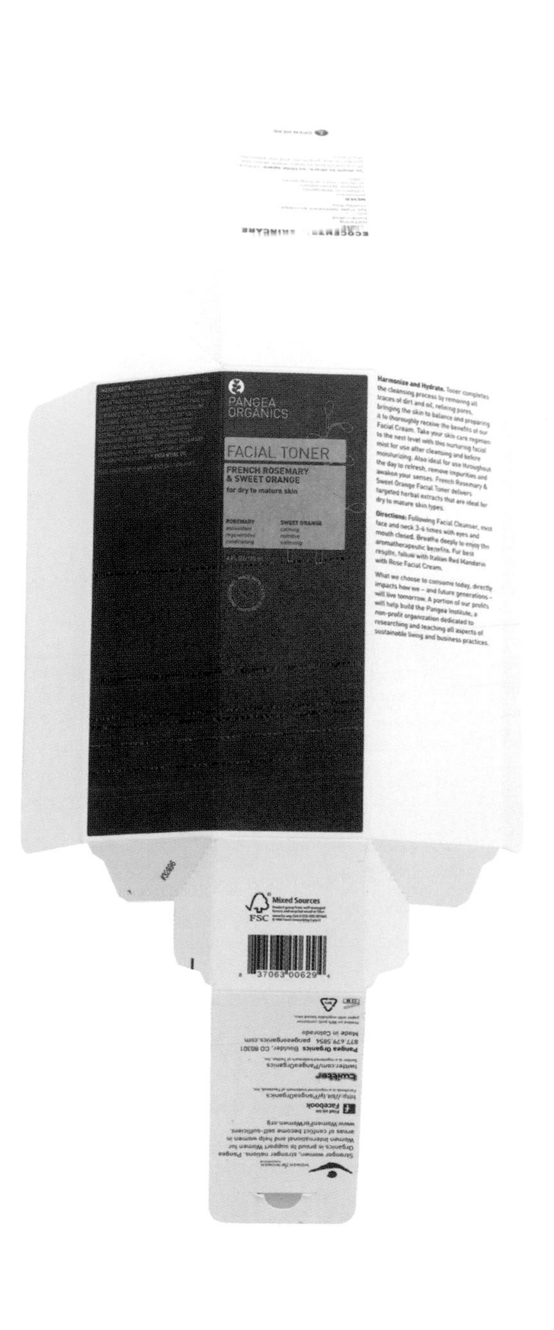

its environmental footprint. The paper used is called WindPower 80 Recycled Board, an uncoated paperboard made of 80 percent post-consumer recycled fiber. It is 100 percent wind-energy produced, carbon-neutral, and ECF (elemental chlorine-free) and PCF (processed chlorine-free). It uses no adhesives and is printed with a vegetable-based ink in two colors on both sides by an FSC-certified printer in the same US state the products are produced in. The packaging is a basic, effective solution that stands the test of time and can serve as a guide for future solutions.

A quest for intelligent, sculpted solutions results in a sophisticated, glueless box that is entirely different from typical packaging designs.

FILTER LIGHT

DESIGNER
Bijl Architecture
www.bijlarchitecture.com.au

DESIGNER BIOGRAPHY
Based in Sydney, Australia, Bijl Architecture is an architecture and design practice led by director and architect Melonie Bayl-Smith and established in 2002. The firm's residential, interiors, and commercial work is underpinned by the practice's involvement with teaching and research. Its approach encompasses aesthetics, functionality, economy, and the total environment, while advocating flexibility and longevity as key to a sustainable architecture and design.

MATERIAL
Corrugated cardboard

MATERIAL PROPERTIES
Lightweight, Recyclable, Simplification, User Experience, Primary Packaging

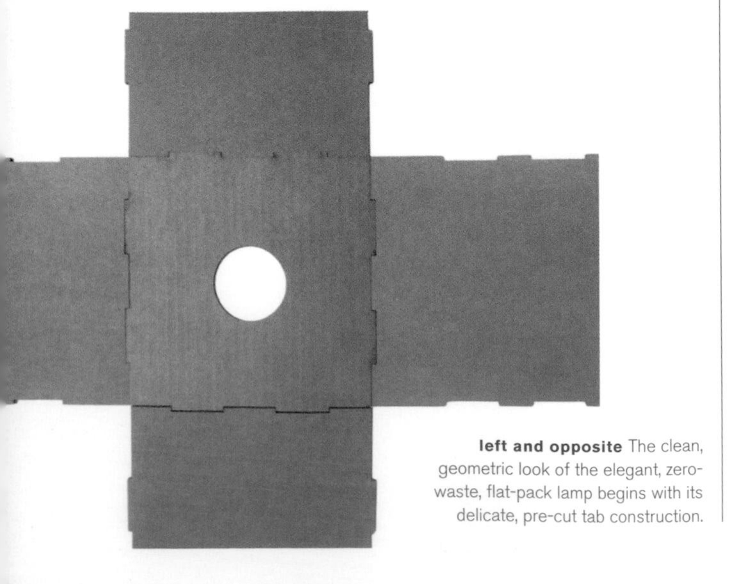

left and opposite The clean, geometric look of the elegant, zero-waste, flat-pack lamp begins with its delicate, pre-cut tab construction.

Whether sculpturally complex or sleekly minimal, today's feats of packaging ingenuity push the field in new directions, upending the traditional relationship between contents and container. The designer of the Filter light, Melonie Bayl-Smith, is an architect, illustrating the way practitioners from all fields now address diverse practical design needs.

Australian-based Bijl Architecture originally designed the Filter light for the Lightcycle design competition, recognizing an opportunity to "investigate the way we interact with packaging, with products and the life cycle of the things that we consume," as Bayl-Smith explains.

Since objects that can be quickly and easily assembled often require fewer materials and are more efficient to ship, she turned her attention to "flat-pack," translating the benefits of prefab architecture into packaging. With the aid of advanced modeling techniques, the firm arrived at a versatile flat-pack lamp whose shade is constructed from its own packaging, a zero-waste solution. For the prototype, a delicate pre-cut-tab construction made from cardboard and acrylic delivers a planar structure and symmetrical silhouette, over which a semi-opaque sleeve can be raised or lowered to adjust the quantity of light. The light can also be rendered in wood and metal as a floor lamp.

MATERIAL INSIGHT

Designed for cardboard and acrylic applications, this customizable lighting fixture takes the methodology of flat-pack and applies it to both product and packaging. Precision-cutting of the cardboard shapes and the use of pre-cut tabs result in minimal assembly with no need for glues or mechanical fixtures. This also allows for easy adjustment, varying the amount of shading. Indeed, there is no correct way to construct the light, with multiple variations possible.

"Y SPINE" FOR SAYL CHAIR

DESIGNER
Herman Miller Inc.
www.hermanmiller.com

DESIGNER BIOGRAPHY
Lead packaging engineer: Cindy Doman; lead production supervisor: Terry Lopez; lead facilitator: Carlos Argueta; lead continuous improvement: Jeff Geurink; lead material handler: Brian VerStrate. Herman Miller's award-winning designs, inventive technologies, and strategic services help people do great things and organizations perform at their best. The company's furniture designs can be found in the permanent collections of museums worldwide, while their innovative business practices and commitment to social responsibility have established their reputation as a global leader. For the "Y spine" Herman Miller's cross-functional team combines manufacturing operations, distribution, part-supplier, and packaging-supplier expertise to deliver the optimal packaging solution.

MANUFACTURER
Cannon Equipment, Kentwood Packaging Corporation
www.cannonequipment.com
www.kpcpackaging.com

MATERIAL
Injection-molded plastic

MATERIAL PROPERTIES
Lightweight, Second-Life, Durable, Low Energy Consumption

Some of the most significant packaging innovations happen far from the retail shelf, in manufacturing plants and on assembly lines. In the competitive world of upscale office furnishings, minimizing the journey a part must travel and the number of times it must be handled to arrive in the correct position for easy assembly represents a sustainable business practice. In fact, reducing time and minimizing waste of any kind is at the heart of the production process.

At Herman Miller, the decision to switch from expendable to reusable packaging for a single element—the Y-shaped spine on the Sayl office chair—delivered savings at multiple levels without jeopardizing quality. "We are always looking for ways to reduce waste, whether it is packaging material or extra labor needed to repack parts for efficient presentation to the assembly lines," explains packaging engineer Cindy Doman.

The key was the introduction of a returnable filler with crosslink foam for the shipping and handling of a crucial part made from glass-filled nylon that cannot be scratched or scuffed. This critical piece is now supplied to Herman Miller in racks that hold 180 identical units at a time, ready for assembly as soon as the racks arrive. Once depleted, the packaging is ready to be filled up again. No more corrugated boxes, and no more time spent dismantling and discarding them. This step alone eliminated 11,180 kilograms (24,645 pounds) of packing material annually, revealing how big changes can begin with small actions.

The chair's form emphasizes the designer's role as rigorous editor by stripping away all extraneous components, an approach that extends to the logistics of assembly.

MATERIAL INSIGHT

This unique packaging solution created specifically for this component of the Sayl chair represents only a fraction of the overall packaging system designed by Herman Miller for low environmental impact and zero waste. Based on the principles of Toyota's "lean manufacturing system" (originally known as "just-in-time production") and using the US Environmental Protection Agency's Design for Environment (DfE) practices, the company has approached zero waste through the use of returnable packaging that is shipped back to the supplier for repeat uses. This increases costs in terms of materials and the creation of part-specific jigs, braces, and supports at the start, but these are amortized over the life of this highly durable packaging. The Sayl packaging holds multiple "Y spines" in a configuration that can be easily filled, shipped, and removed for assembly of the chair. No additional protective cardboard, foam, or film is needed, and the empty jig is returned to the supplier. This type of approach requires that the supplier also be consulted to ensure that the solution is effective for the entire life cycle of the packaging, and requires a high level of control over all aspects of production. This process lends itself to products that are unlikely to have design changes from year to year (such as office furniture) and to more localized and committed vendors.

In the highly competitive world of contract furnishings, any production refinement that improves the assembly process, though it may be invisible to the consumer, is an innovation to be celebrated.

VIRGIN ATLANTIC ECONOMY

DESIGNERS
Virgin Atlantic: David Gadd, Nathan Sparshott
MAP: Jon Marshall, William Howe, Scott Barwick
www.mapprojectoffice.com

DESIGNER BIOGRAPHY
Founded in London in 2012, MAP strives to deliver a more intelligent, end-to-end approach to industrial design. Part of a family of world-class design companies, alongside *Barber & Osgerby* and *Universal Design Studio*, this multidisciplinary team of industrial designers, strategists, and innovation experts is united by a common approach: a love of craft and of detail, materiality, and innovative design thinking.

CLIENT
Virgin Atlantic Airways
www.virgin-atlantic.com

MATERIAL
Polyester (PET and rPET), styrenes (PS, SAN, and ABS), polypropylene (PP), and thermoplastic elastomers (TPE)

MATERIAL PROPERTIES
Lightweight, Ergonomic, Recycled, Recyclable

Leave it to Virgin Atlantic, a renowned business and design innovator, to prove that order can be seductive. As the realities of climate change have taken hold, the airline industry has faced heavy scrutiny of its CO_2 emissions from fossil fuel. Studies on airline waste reveal a dismal record, with most trash sent to landfills and incinerators. In response, Virgin initiated the three-step "reduce, reuse, recycle" approach: Reduce the amount of waste that is generated, and reuse and recycle as much as possible of what is left.

But they also went a step further. In the air, every gram saved reduces carbon emissions, so Virgin invited MAP, the UK-based industrial design consultancy known for its love of craft and detail, to collaborate on a solution to reduce weight and at the same time to improve the entire food delivery experience from end to end.

Following extensive research and testing, myriad smart improvements add up to a refined yet economical system. The redesigned trays, for example, consume less space and are lined with non-slip pads to keep containers in place. Lively graphics plus purple cups and cutlery add a vibrant dynamism to the overall presentation. And a new hot-beverage pot with an ergonomic handle features a lid with an arrow that points to the type of contents labeled on the rim. The overall focus on user experience is consistent with Virgin's signature style and contributes to its exceptional brand loyalty.

MATERIAL INSIGHT

A wide range of commodity and more innovative materials was utilized for the system. Plastics such as polyester (PET and rPET), styrenes (PS, SAN, and ABS), and polypropylene (PP) were combined with recycled plastics as well as soft-touch over-molded thermoplastic elastomers (TPE) for touch points on the tray. Everything but the tray and the salad pot was designed as a consumable. Items were produced with lightweight synthetic materials and as individual components that did not combine materials, so that they could be effectively recycled. The significant reduction in weight (129 kilos [284 pounds] per aircraft) was achieved through design, but also through careful use of low-density polymers for each of the pieces.

The humble meal tray has been transformed through a subtle and rigorous design process that included flying with the cabin crew to learn firsthand how flight attendants and passengers experience dining in flight.

A carefully engineered collection of tableware reduces storage requirements, carbon emissions, and costs while improving the overall tidiness of the delivery.

GRO-ABLES SEED POD

DESIGNER
Group 4
www.groupfour.com

DESIGNER BIOGRAPHY
Group 4 is a research-based design consultancy that has helped develop hundreds of products, packages, and brand communication strategies for a wide range of clients, including Sherwin-Williams, Nestlé, Lenovo, Electrolux, UPS, and Eli Lilly. Research, design, and engineering teams worked closely together throughout the development process. Director of product structural design Robert Bruno led the design teams' effort in developing the form factor and functionality aspects of Gro-Ables. Under his leadership, Group 4 projects have been recognized by IDSA/IDEA, Good Design, and DuPont Packaging, among others. Matthew Phillips, director of engineering, led the materials evaluation and innovation, and helped define production and assembly processes with Scotts Miracle-Gro. He possesses numerous utility and design patents through his work for such clients as Pitney Bowes, Reebok, Dow-Corning, and Seek Thermal.

MANUFACTURER
Scotts Miracle-Gro
www.miraclegro.com

MATERIAL
Natural cellulose pulped fiber

MATERIAL PROPERTIES
Shelf Impact, Low Toxicity,
User Experience,
Primary Packaging

The intuitive design of the Gro-ables Seed Pod emphasizes the role of the gardener rather than the garden, to inspire more people to get their hands dirty.

While the Scotts Miracle-Gro Company is renowned for its lawn-care products and liquid fertilizer, before the two brands merged they shared a common heritage in selling seeds and plants. The company has continually experimented with new products, especially ready-to-use formulations and reduced packaging. As the desire for locally sourced food has fueled the farm-to-table movement, along with the proliferation of "microfarms" in cities for growing food and increasing biodiversity, the climate has been especially ripe for an innovation in seed gardening.

Even for experienced gardeners, however, seed gardening is hardly foolproof, often involving guesswork with disappointing results. So Scotts Miracle-Gro's innovation team embarked on a collaboration with Group 4 that leveraged research and product and packaging design to resolve materials and ease-of-use challenges.

The result is the Gro-ables Seed Pod, a biodegradable container that packages seed, soil, and fertilizer in an ideal growing environment. With its flat top and tapering sides that mimic an acorn, the design emphasizes the physical properties of earth, thereby underscoring an oft-repeated rule of successful gardening—put a one-dollar plant in a ten-dollar hole. The pointed tip nestles easily into the soil, while the wide brim marks the optimal planting depth. Through a lengthy iterative research and design process, this ingenious, aesthetically simple solution unlocks the potential that lies in a tiny seed, allowing edible plants and green thumbs to flourish.

MATERIAL INSIGHT

The form, aesthetics and composition of this packaging all contribute to its effectiveness as a plantable product. The "tri-acorn" shape can easily be pushed into the soil to the correct depth. The color and fibrous texture of the molded

container reassures gardeners that it can be safely left in the ground, surrounding the growing plant. In addition, the natural cellulose fiber and binder that constitutes the container itself degrades into biomass during the growing season, also fertilizing the shoot. As an intuitive, sustainable approach to innovation, this packaging solution emulates nature's cyclic processes and dramatically simplifies the conditions essential to healthy plant growth. The distinct advantage that this system offers is that the container pod is compatible with the surrounding soil, breaking down quickly into usable nutrients for the plant.

The simplicity of the pod design, where packaging and product are one, supports the increase in urban vegetable growing, which allows city-dwellers to experience freshly grown seasonal produce while offering a visceral reminder of where food comes from.

LASER-ETCHED PRODUCE

DESIGNER/MANUFACTURER
Laser Food
www.laserfood.es

DESIGNER BIOGRAPHY
Laser Food 2007, SL is a Spanish company based in Valencia that has created a unique system of labeling fruits and vegetables by means of a laser light.
The technology was developed with the help of the European Commission within the Eco Innovation program. Laser Food machines are present in several countries and the company is growing steadily.

CLIENT
Various

MATERIAL
Carbon dioxide (CO_2) laser

MATERIAL PROPERTIES
Lightweight, Non-Toxic, Simplification, User Experience

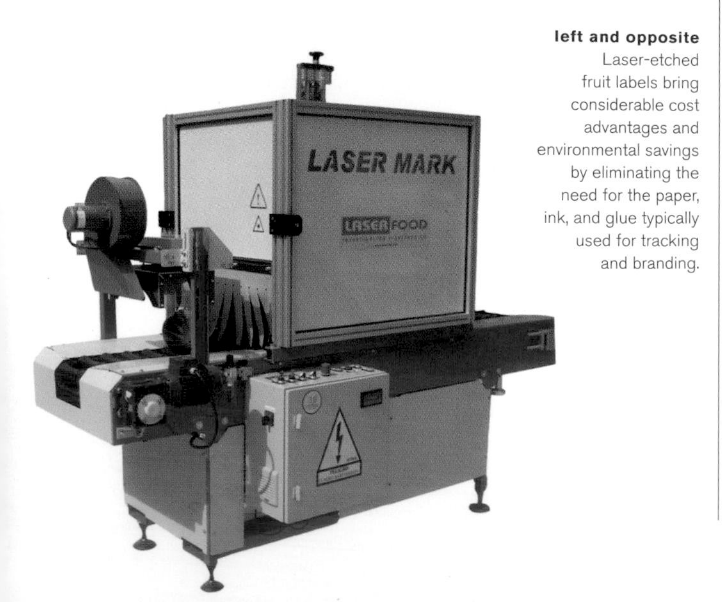

left and opposite Laser-etched fruit labels bring considerable cost advantages and environmental savings by eliminating the need for the paper, ink, and glue typically used for tracking and branding.

Whether the source is a family-owned farm that grows heirloom varieties or a vast orchard that is part of the world's tropical fruit production, tracing the journey that a soft-skinned, edible product makes from farm to table is no easy task. Typically the process involves tracking a minuscule sticker containing a bar chart no larger than an infant's thumb. But as concern grows regarding the origins of the food we buy, from the use of pesticides to labor practices, efficiently tracing fruits and vegetables by the piece is of value to producers as well as consumers.

Eliminating stickers altogether is the goal of Laser Food. Through a collaboration with researchers from the University of Valencia, the Spanish-based company developed a laser-etching process that brands the surface without penetrating the skin. The first step de-pigments the exterior skin, then a contrast liquid containing iron oxides and hydroxides is applied, leaving a "fruit tattoo" that does away with the paper, inks, and glues needed for stickers. The marking costs less than a penny, takes a fraction of a second to execute, and can be customized for proprietary branding and seasonal messaging.

MATERIAL INSIGHT

Using a focused low-energy carbon dioxide laser that works in the infrared region of the spectrum, markings can be made on the rinds, skins, and shells of fruits, vegetables, and nuts by burning the top few hundred microns of the surface. A gas-discharge laser is used, which passes a current through the CO_2 (actually CO_2, N_2, and He) to create an ionized discharge, which is focused using mirrors. The contrast between the laser marking and the skin of the fruit or vegetable can be enhanced by coating the produce with iron oxides and hydroxides, though these are currently recommended only for produce where the skin is not eaten (such as citrus fruits, melons, and watermelons). According to research, the marking does not noticeably impact shelf life and is not an accelerator of bacterial attack. It ensures accurate inventory control (no lost or mixed-up stickers), enables branding, and allows nutritional and environmental impact information to be included, all with the use of zero physical material.

EDIBLE FOOD NESTS

DESIGNERS
Dr. Diane Leclair Bisson and Vito Gionatan Lassandro
www.edibleproject.com

DESIGNER BIOGRAPHY
Based in Montreal, Canada, Dr. Diane Leclair Bisson holds advanced degrees in Anthropology, Museology, and History of Design. Her design projects merge cultural research and experimentation with materials and technology to give products a socially and environmentally meaningful value. Since 2000, she has been exploring the world of sustainable food practice and nutrition, launching the pioneering Edible Container Project to reduce packaging waste and publishing *Edible: Food as Material* (Les Éditions du Passage, 2009).

CLIENT
Salone del Mobile, Milan

MATERIAL
Tomato, agar agar, and juices or vinaigrettes

MATERIAL PROPERTIES
Rapidly Renewable (Resources), Low Toxicity, User Experience

Some have the dynamic quality of kinetic sculptures. Others are impressive for their delicate details. Diane Bisson's lively array of edible food nests draws on a range of motifs—a Chinese takeout container (inanimate), bird nests (animal), and interlocking hands (anthropomorphic)—that refer to culinary traditions, imbuing her consumable tableware with a layer of cultural significance that is not unexpected from an industrial designer who is also an anthropologist.

Although the impression is of something far-fetched, Bisson's designs speak to the elegance of the creative mind as much as the more pressing initiative that inspired them: Taste No Waste. "In the beginning of culinary history, edible food containers were always born directly from their function: to hold food, absorb sauces, and complement the meal," she explains. "Today, they address needs that are no longer merely economic or functional. They speak to environmental sustainability and aesthetics."

Seeking to evolve ideas about how we interact with foods and their containers, Bisson renders potato and tomato pulp in textures from soft to crunchy to jelly-like. Some nests fit easily in the palm of the hand, while others are large enough to be shared. "The edible containers call for a new aesthetic, a new typology of shapes and food handling that invites people to discover and explore sustainable eating practices."

left and opposite Edible food nests explore the relationship between container and contents, turning mundane culinary ingredients into spectacular handheld sculptures.

MATERIAL INSIGHT

The concept of using edible or cookable materials as a packaging system for food is not new, but rarely has it been approached with such consideration for the aesthetic appeal of the packaging. The two systems shown here, one rigid and somewhat friable, the other jelly-like, are composed of tomato for the former and a solution of agar agar and diverse liquids such as juices or vinaigrettes for the latter. In very much the same way as the foods they will protect, these packages were developed using cooking methods, but those more akin to industrialized processing, where adherence to equipment, molding processes, and final protective treatments was considered in their production. Even the pigmentation of the packaging was considered in terms of color stability, thermal performance, and, of course, nutritional value.

opposite and below Edible food nests force us to see beyond typical expectations of consumable containers, introducing an extra degree of technical innovation and aesthetic expression akin to gastronomy.

"THE DISAPPEARING PACKAGE" TIDE PODS

DESIGNER
Aaron Mickelson, Pratt Institute
www.disappearingpackage.com

DESIGNER BIOGRAPHY
After earning an MSc in Packaging Design from the Pratt Institute, New York, where he received a Pratt Circle Award, Aaron Mickelson worked for packaging agencies in New York City and Chicago before launching his own company, Alocato. His designs have appeared in *Fast Company*, *Wired* and *Food & Beverage Packaging* magazines.

MATERIAL
Polyvinyl alcohol (PVOH)

MATERIAL PROPERTIES
Lightweight, Shelf Impact, User Experience, Primary Packaging, Secondary Packaging

A dissolvable exterior packaging transforms the nature of the product by eliminating mess and waste as well as the bulk of large plastic containers.

Reduction can be the wellspring of new ideas. "I was considering a blue-sky refresh of commodity packaging. On a whim, I thought about applying the functions of packaging to the product itself and was struck by the green potential of an idea that could work across several product types." So began Aaron Mickelson's effort to show that elimination of packaging does not require a paradigm shift and might even generate new possibilities.

With zero waste as the goal, Mickelson, then a master's student at Pratt Institute, began by methodically walking store aisles in search of the needless and superfluous. "Before proposing a new solution, I audited the existing packaging to discern each brand voice and identify consumer expectations." He chose five household products that merited eco-friendly makeovers.

For Tide Pods, he stitched together a page of tablets into a single perforated sheet, then printed brand messaging on the back in soap-soluble ink and tightly rolled them up for store display. At home, the Pods are torn off one at a time until none is left; the product dissolves in the machine, sparing the landfill nearly 660 kilos (1,450 pounds) of packaging waste per delivery truck.

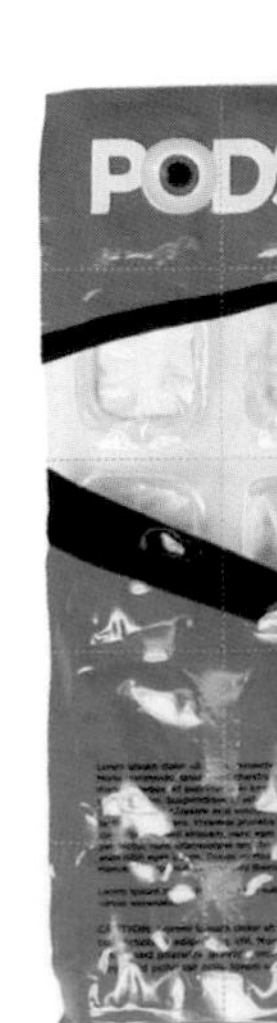

Mickelson's striving to realize a packaging-free future, however, revealed that the toughest links in the chain are packaging manufacturers and material suppliers, whose resistance to innovation is not without its challenges.

MATERIAL INSIGHT

The ultimate in efficient zero-waste packaging must be that which disappears after its useful life protecting the product. The use of plastic films made from polyvinyl alcohol (PVOH)

in products such as Tide laundry Pods, where the film containing the liquid dissolves in the water along with the detergent, only gets halfway there, since there is always an external box or bag to contain the Pods. This conceptual version of the same product, however, imagines the use of that film as a "tear-off" strip, with the branding, usage instructions, and other necessary information printed right on the product in water-soluble ink. PVOH film prints and takes ink well—it is used as a carrier for hydrographic printing—but the inks need to be dissolvable and generally non-toxic. All that is needed to ship the product is a water-resistant thin film to protect it against high humidity conditions and wet hands—this is about as close to zero waste as you can get.

Like garments packed for a trip, Tide Pods are rolled up for shipping to maximize space, then unfurled for use.

DISSOLVE

DESIGNER
Simon Laliberté
www.atelierbangbang.ca

DESIGNER BIOGRAPHY
Simon Laliberté heads Atelier BangBang, a multidisciplinary design studio and screenprinting workshop based in Montreal, Canada. After graduating from the Université du Québec à Montréal (UQAM) in 2012, he worked for Renzo Designers, Chez Valois, *Urbania* magazine, and Bleublancrouge, and at the same time launched his own business. His designs have been recognized with a Gold Pentaward and exhibited in the Biennale Internationale Design Saint-Etienne, France, and he has won three Grafika awards since 2012.

MATERIAL
Polyvinyl alcohol (PVOH), cellulose pulp, and soy-based inks

MATERIAL PROPERTIES
Printable, User Experience, Secondary Packaging

There is something innately logical about a toothbrush package that dissolves when wet. While a design student at the Université du Québec à Montréal, Simon Laliberté drew on the ability of water to slowly erode almost any material as part of an assignment to upgrade the packaging of an existing product with a design that disappears. In the case of a toothbrush, his solution also references the fate of the product itself—billions of toothbrushes are thrown away each year as part of recommended oral hygiene.

While toothbrush design has continued to evolve, from compostable to solar-powered to bristle-free options in every conceivable size, shape, and stiffness, the packaging has largely remained the same, an endeavor of near-impenetrable transparency dominated by hard plastic with bright, color-coded graphics for easy selection.

With "Dissolve," all that changes. In a gesture both poetic and propagandistic, the biodegradable box simply disintegrates when it is run underneath tap water. What is left in its absence is the suggestion of a new set of environmental possibilities for any packaging commonly used in the kitchen, bathroom, or laundry.

Elegant black-and-white graphics strike a perfect note for a product that needs to stand out in a sea of technicolor on the drugstore shelf. As the user moves from one end of the box to the other, the three-sided carton shifts from a vertical to a horizontal seam, mimicking the physical mechanics of brushing all one's teeth.

Dissolve's classic black-and-white graphics give it a timeless quality, while its sleek triangular shape looks entirely new and different for a toothbrush package.

MATERIAL INSIGHT

The easy cold-water dissolvability of this packaging is achieved by the use of polyvinyl alcohol (PVOH) as a major constituent of the paperboard. PVOH is a clear plastic, and in this case was used to bind the cellulose of the paper. It is cold-water soluble, causing the package to very quickly

break down and "dissolve" (in less than ten seconds) so that it can be washed down the drain. The chemistry of PVOH can also be engineered to break down only in hot water, for applications where cold-water integrity is required. The inks used are soy-based and also water-soluble, and the package is heat-sealed rather than glued, ensuring that there are no non-dissolvable components. The cellulose pulp is removed from the water safely at treatment plants, much the same as toilet tissue, though it is classed as generally non-toxic and can be safely ingested by fish if dissolved in open seawater.

NIKE AIR MAX CUSHION SHOEBOX

DESIGNER
Scholz & Friends
www.s-f.com

DESIGNER BIOGRAPHY
With offices in Hamburg, Berlin, and Dusseldorf, Scholz & Friends is ranked among the top three creative advertising agencies in the German-speaking world. Now part of the WPP network of marketing communications services, it was founded in 1981 by Jürgen Scholz, Uwe Lang, and Michael Menzel.

MATERIAL
Polyvinyl chloride (PVC)

MATERIAL PROPERTIES
Lightweight, Shelf Impact, Durable, Second-Life

below and opposite A clear inflatable pillow challenges conventional notions of packaging, forcing us to consider a limitless space seemingly without intervention as part of the language of protective containers.

The celebrated "air bubble" in the sole of the Nike Air Max shoe stands as one of the sportswear giant's most significant innovations. Inspired by the exposed functional elements of the Centre Georges Pompidou in Paris, vice president of design and special projects Tinker Hatfield sought to showcase the airbag cushioning in the heel of the shoe by installing a visible window that would allow consumers to see inside.

This core idea is at the heart of a prototype packaging solution conceived by Scholz & Friends. Inside a taut, transparent pillow, a pair of sneakers appears to float, offering an arresting study in weightlessness. The inflated design is an ingenious response to the dual demand for a protective shipping container that can both display the shoe from every angle and encase the product in the same clear, futuristic bubble that its own form encloses.

As with so many Nike designs, the packaging confounds our typical expectations by altering how we usually interact with a product meant to be worn. The aesthetically minimal solution transforms the banality of a cardboard shoebox into something buoyant and uplifting.

While air-filled forms and inflatable objects can be traced back to eighteenth-century hot-air balloons, the sense of wonder they inspire remains, and the customer is taken momentarily to a place that operates according to different laws of gravity.

MATERIAL INSIGHT

The most important "material" in this concept is actually air, in that it is the sealed air that ensures the shape and cushioning of the packaging. The right material to contain that air is essential, however. Applying the simple functionality of bubble wrap to create one large bubble necessitates a change in material type to polyvinyl chloride (PVC). This commodity resin, extruded into a film, is the

ideal material for this application, as it is highly transparent, can be heat-welded to seal the edges, keeps its shape, and is an excellent air barrier, meaning that it will not deflate over time. Although PVC has lost favor in many companies owing to concerns about chemical toxicity (the concept developers could also have used a suitable clear thermoplastic polyurethane [TPU] film with similar results), recent developments in plasticizers—the additive that makes PVC soft—that are phthalate-free mean that it may yet be a source of excellent packaging again.

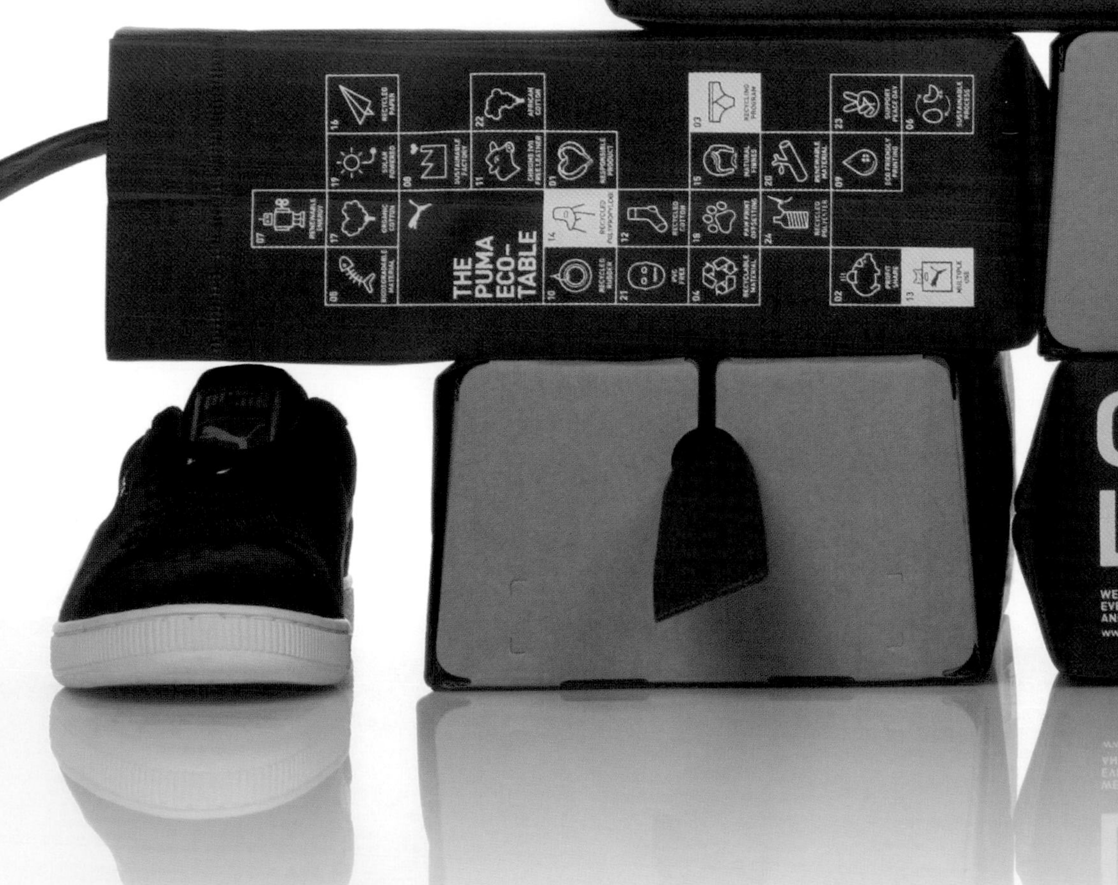
CLEVER
LITTLE BAG
WELL IT'S SMARTER THAN AN OLD FASHIONED SHOEBOX BECAUSE IT USES 65% LESS PAPER. EVEN BETTER, IT MEANS YOU DON'T NEED AN EXTRA CARRIER BAG AND YOU CAN USE IT OVER AND OVER AGAIN. CLEVER HUH? FOLLOW THE PUMA ECO-TABLE. REUSE THIS BAG.
www.puma.com/cleverlittlebag
THE PUMA ECO-TABLE

VER
LE BAG
HIONED SHOEBOX BECAUSE IT USES 65% LESS PAPER.
ED AN EXTRA CARRIER BAG AND YOU CAN USE IT OVER
W THE PUMA ECO-TABLE. REUSE THIS BAG.

BUELL

THIS CRATE IS ENGINEERED TO BECOME A WORK BENCH. Engineering isn't something you turn on and off. If you're an engineer, it's just there, in your head, all the time. So when we got through engineering the bike, we kept right on engineering, onto the crate it came in. If we owned the delivery truck that the crate was loaded onto, we would have engineered that as well. But we don't. So we didn't.
Buell. Ruthless Engineering.
BUELL

CHAPTER 2
FUNCTIONAL FORMS

All packaging needs to serve the purpose of protecting its contents, but there are examples where the packaging offers a distinct additional function that enhances its status above just simple protection: It is these innovations that we highlight in our chapter on functional forms.

This section presents examples where the innovation is in the creation of a new form or system to solve a packaging problem. Although the resulting forms may offer their own particular aesthetic, this is considered secondary to the engineering solutions that they provide. On a basic level, the majority of all packaging fits this category, though consumers rarely see it, as it tends to be used for the shipping of parts from factory to factory, or as larger-format "secondary packaging" (external to the main container) that enables the primary to be shipped in large quantities without damage (think of a large cardboard shipping box containing forty smaller decorated store-ready boxes that in turn package the product). Owing to cost constraints and sustainability considerations more and more packaging is solving these challenges through the use of "functional forms."

Forms can reduce overall shipping volume or, as in the case of the AidPod, piggyback on existing delivery systems to provide much-needed wider dissemination. They can aid recycling, especially when considering the very common "hybrid" containers that use multiple materials to enhance performance. They are also able to create new ways to experience products, such as the Coffeebrewer, which through multilayering of materials enables easy and safe filtering of scalding hot coffee.

opposite The convertible aspect of Buell's shipping-crate/workbench speaks to their target market—expert riders who prefer to assemble and repair their own motorcycles—and eliminates waste from discarded packing crates.

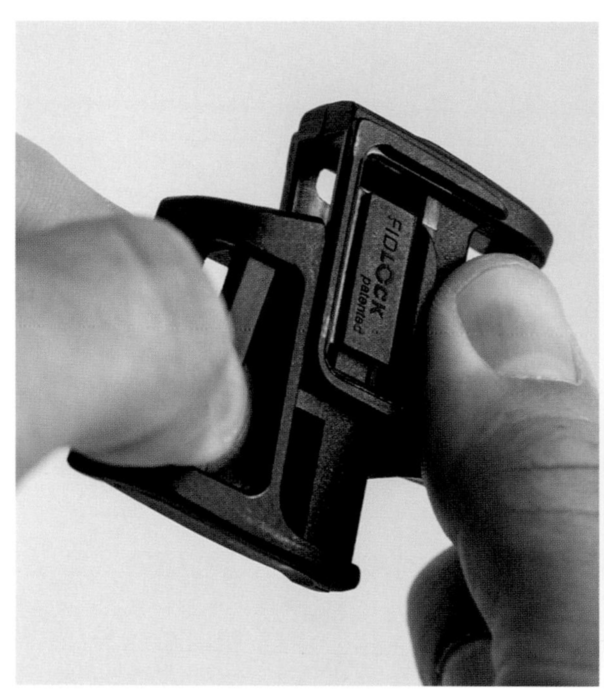

The Fidlock series of fasteners uses complex molded male and female parts along with internally located permanent magnets to create lock and release mechanisms that are secure, intuitive, and able to be operated with one hand.

Materials play an essential part in this functionality, often as a secondary role to a new process, and other times as the driver of the innovation itself. For most packaging requirements, the need to keep costs to a minimum has meant that much innovation in materials has tended to be within value-based lightweight materials such as commodity plastics, paperboard, cardboard, and some metal foils. This constraint in materials has forced significant advancement in these areas in the last decade or so, with new fiber developments for papers, enhanced corrugation for cardboard, improved multilayers of metal, polymer, and paper foils, as well as better performance of polyolefins (the polyethylene PE and polypropylene PP family of thermoplastics) as well as polyesters (PET, PET-G) and to a certain extent polyvinyl chloride (PVC). It is worth noting, however, that the use of the latter versatile plastic has precipitously declined in this period owing to concerns (not always justified) around toxicity of some of the additives used (plasticizers called phthalates) and following end-of-life if burned for disposal. Desire for more sustainable solutions has also led to a rise in bioplastics, the addition of bio-based raw materials to plastics, and improved recycling. Most current bioplastics in packaging are based on bacterial interaction with starches such as polylactic acid (PLA), and the polyhydroxyalkanoates PHA, PHB and PHBV. Basic starches that are by-products of the food industry are also being widely used as additives to oil-based commodity plastics to reduce petrochemical use and to enhance potential biodegradability and compostability. Recyclability is also being made easier with greater use of a single polymer in multiple parts such as PP, as well as "pull apart" solutions for hybrid packages.

The dividing line between molded paper pulp, such as seen for egg containers, and injection-molded plastics has become more blurred, with injection-molded parts being created using paper fibers with little or no plastic, and also parts that might look like molded pulp but are in fact predominantly plastic or bioplastic. Wood or paper fibers (cellulose-based) are being modified through treatment to bond better to each other, to withstand more aggressive environments (chemical, water, heat, etc.), and even to pass leather standard tests.

When a single layer of plastics or paper cannot give sufficient performance, multiple layers of different materials are able to provide a synergistic effect with little increase in weight or cost. Developed predominantly for food preservation but now finding wider application in medical-, pharmaceutical-, and consumer-product packaging, these multilayer film structures are able to block or allow the passage of gases such as oxygen, carbon dioxide, nitrogen, and ethylene, protect against UV, and make the surface more printable, more reflective, softer, quieter, or even electronically interactive. Individual polymer layers can be as thin as 0.5 microns (plastic wrap, or cling film, is 12 microns; a human hair ranges from 20 to 180 microns), and metal foils at 0.2 microns are routinely used, becoming light-translucent at this thickness. Anything less than half a micron will typically need to be supported by a thicker "strength layer," which may not have any function beyond acting as a supporting substrate for the functional layers above and below.

These material innovations allow designers and engineers greater freedom and creativity to develop forms and solutions to fulfill existing packaging needs, but also to offer completely new ideas that push packaging design into new areas, even influencing the design of the products they protect. Though not always the most attractive packaging, functional forms are often the most revolutionary and thought-provoking.

Industrial origami uses metal's combination of plastic deformation—the ability to be bent into shape—and cutting precision to create structural parts simply by folding along pre-scored lines.

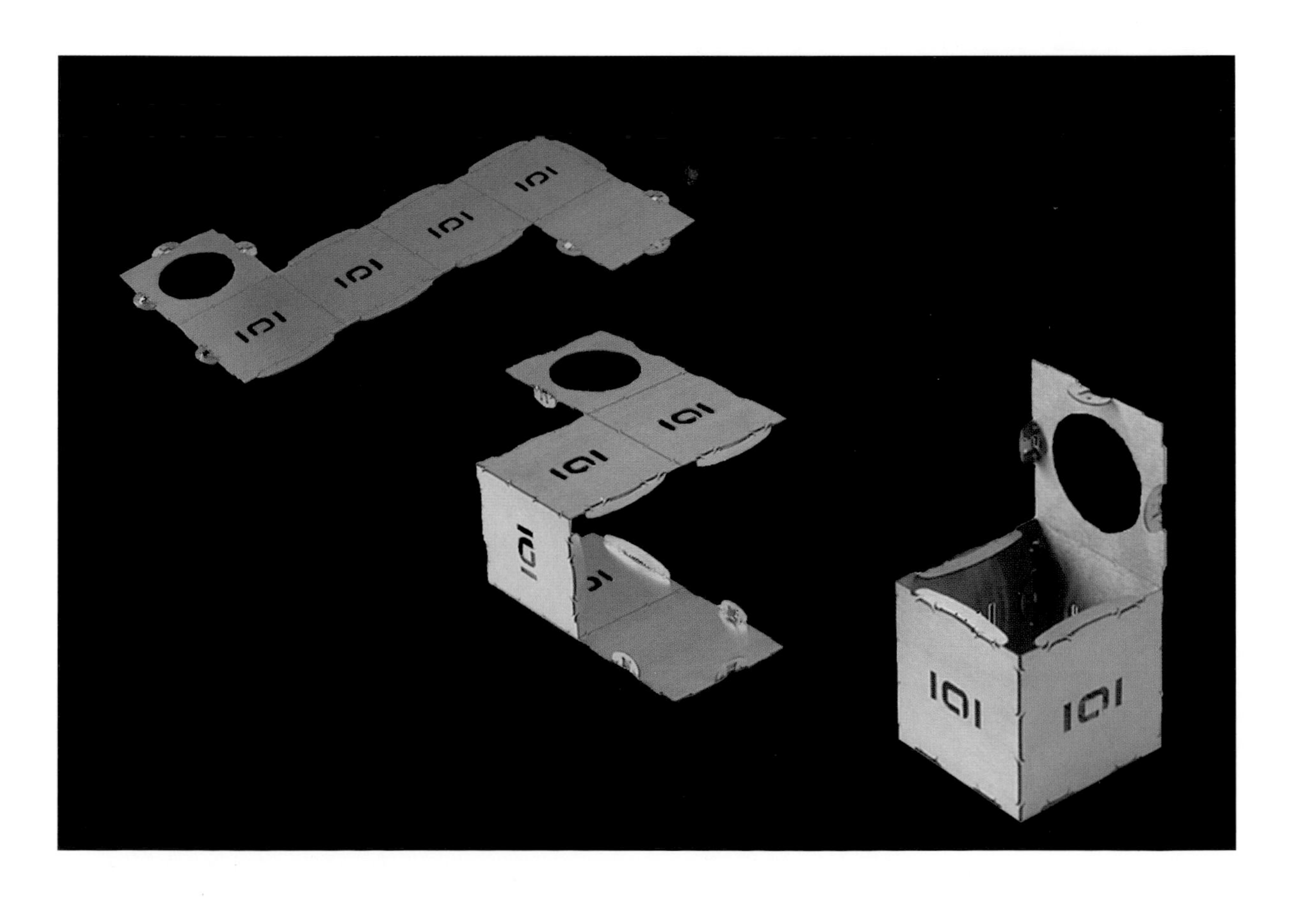

KMS MESSING CREME "HAIR PLAY"

DESIGNERS
Tube design: Kao Germany GmbH
Material development: Johann Beil, lead packaging engineer, Linhardt
www.linhardt.com

DESIGNER BIOGRAPHY
Born in Bavaria, Germany, and trained as an industrial designer, Johann Beil has helped shape the design process at Linhardt since 1988. He established and leads the Linhardt research and development department, where his primary focus is the development of packaging systems that are aligned with strategic corporate planning, especially when it involves proprietary design solutions.

CLIENT
Kao Germany GmbH

MANUFACTURER
Linhardt GmbH & Co. KG
www.linhardt.com

MATERIAL
Thin filmstrip

MATERIAL PROPERTIES
Flexible, Multi-Purpose, Printable

The Hair Play tube's sleek, seamless, yet pliable exterior belies the advanced-composite technology underneath.

Beauty trends come and go, but in recent years a less-is-more approach has prevailed in hairstyles. For stylists, it has made dirty hair not only acceptable but desirable. Achieving the right "dirty," however, requires the right product.

At first glance there may not appear to be much that would lure the stylist seeking that perfect lived-in look to choose KMS California's Hair Play Messing Creme over the captivating promises and elegant packaging of competing products, but below the surface is another story. An innovative tube designed in multiple layers by German packaging innovator Linhardt exclusively for KMS boosts the product's shelf presence while opening up new functional and graphic capabilities. With Linhardt's Multiflex tubes, different types of barrier layers inside the tube can be customized according to the product formulation, ideal not only for health and beauty but also for food, pharmaceuticals, and household products. An invisible seam allows for 360-degree printing on a wide range of laminates and gives the package an extra-sleek appearance, an attribute well suited to an industry that correlates product sales with smooth, glossy locks—even when "dirty" hair is the look of the day.

Recognizing that customers will always want a physical experience, Linhardt's in-house team has been a consistent proponent of material exploration as the path to greater technical possibilities, without overlooking the importance of innovations that are simply more pleasing to touch.

MATERIAL INSIGHT

Considered the fourth stage in the evolution of collapsible tubes, after metal, plastic, and then multiple-layered laminate, the innovation behind this new method for sealing flexible tubes, though not immediately obvious, allows designers many more material options for these products. The joint used to create the tube along its side is sealed using a thin filmstrip rather than lap-welding (overlaying the sheets and heat-sealing). Since no heat-sealing is needed, the multilayer does not have to have similar inner and outer layers. This means that a wider range of multilayer foils can be used, including asymmetric foil combinations that use a decorative outer foil (or paper or metal film) with a central barrier foil and an inner layer that is optimal for the filling of the tube. RFID, antimicrobial and other functional layers can also be incorporated, as well as thinner layers that reduce the weight of the packaging by up to 30 percent.

While a beauty product needs to project elegance on the shelf, it must also convey easy extractability, a gesture the Multiflex tube was invented to deliver.

GROWER'S CUP COFFEEBREWER

DESIGNER/MANUFACTURER
Ulrik Skovgaard Rasmussen and Mareks Melecis

DESIGNER BIOGRAPHIES
Ulrik Skovgaard Rasmussen is the inventor of the original Grower's Cup concept and CEO of Grower's Cup, an innovation-driven company intent on changing the way people brew coffee and tea. Mareks Melecis is a freelance graphic and product designer from Latvia, based in Vejle, Denmark. He specializes in the design and development of custom-made paper products. During the past decade, he has assisted Coffeebrewer Nordic with his design expertise.

CLIENT
Grower's Cup
www.growerscup.com

MATERIAL
Multilayer of printable paper and low-density polyethylene (LDPE)

MATERIAL PROPERTIES
Lightweight, Flexible, Low-Tech

The ingenuity of Grower's Cup is the "slow pour" in a single serving, on the go.

Of all the coffee-brewing options available today, perhaps nothing draws out the coffee's fine qualities as much as the pour-over. Slow and low-tech, it is the ideal method for showcasing the complex and varied flavors of beans grown in different parts of the world. A thin, continuous stream of hot water, rather than a saturated dousing, allows the ground coffee to "bloom" and the optimal taste to be achieved. Best done by hand rather than with a machine, it is the coffee equivalent of the tea ceremony.

Enter the Grower's Cup Coffeebrewer, a design that is as much product as package. Recognizing that not everyone is willing to wait for the perfect cup to brew, the Danish company developed the world's first disposable coffee brewer that produces two perfect cups for any situation where its combination of gourmet quality and convenience is appreciated. A waterproof, polyethylene (PE)-coated kraft paper is the outside material, while the interior filter is made from a PE-based mesh. The pouch contains ground beans from one of ten different single-estate coffee farms or co-ops in the exact amount needed to make two pour-over cups in minutes.

As with most designs that appear simple, the countless hours spent cutting and gluing prototypes are not evident. Everything about the package has been thought through to the last detail, a mark of respect for the creative process but also for the consumer.

MATERIAL INSIGHT

The packaging used for this single-use fresh coffee maker has overcome several of the challenges associated with filtering and pouring a hot beverage. The outer bag is a multilayer of printable paper with an inner lining of food-safe low-density polyethylene (LDPE). The lining can withstand hot water temperatures with no loss of performance; it will become

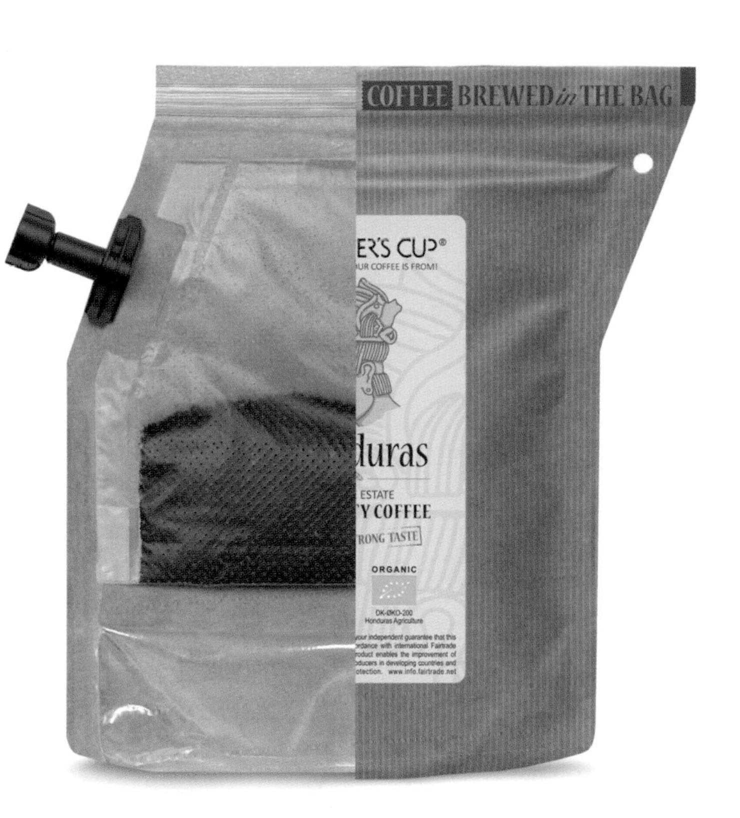

significantly more flexible as it heats up, but the outer paper maintains the necessary rigidity. The inner compartments are also a heat-resistant LDPE, while the upper chamber, which acts as a filter, is a PE-based mesh. The entire construction requires heat-sealing to create the necessary water and coffee flow for brewing and pouring. Because of the materials used, the entire product weighs only 33 grams (1.6 ounces), including the 20 grams (0.9 ounces) of ground coffee. It has a total thickness of only 1 centimetre (0.4 inches) before use. Though it can be safely handled while containing hot water, a recyclable polymer holder can be used for easier pouring.

above left The inner filter compartment is designed to ensure optimal contact between the hot water and the pre-measured coffee grounds.

above The much maligned cup of instant caffeinated brew that can be enjoyed anywhere there is hot water goes from mediocre to exceptional.

below A simple, three-step "open, pour, serve" process makes Grower's Cup Coffeebrewer ideal for a range of users, from students to commuters to outdoor adventurers.

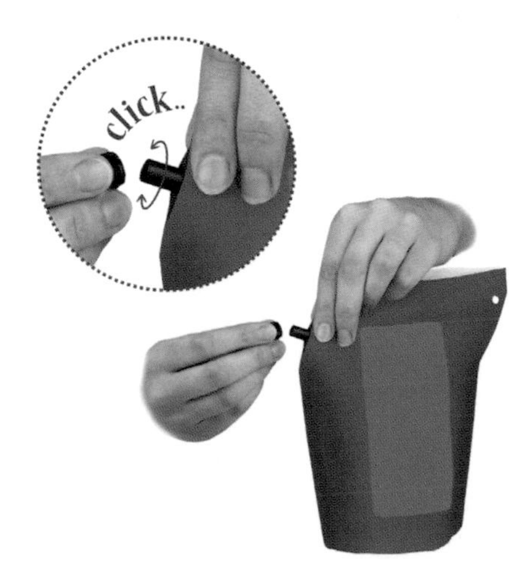

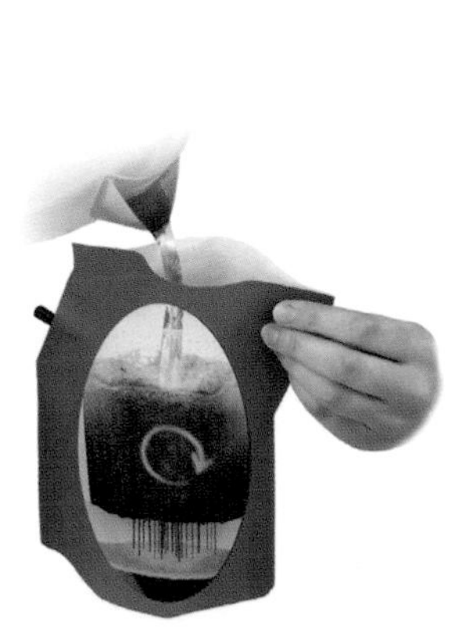

RAPID PACKING CONTAINER

DESIGNERS
Christopher Curro and Henry Wang

DESIGNER BIOGRAPHIES
Chris Curro and Henry Wang met as engineering students at the Cooper Union for the Advancement of Science and Art, New York. Chris, an electrical engineer, believes in mixing disciplines to creative innovative solutions. His particular focus is signal processing and communication theory. Henry is committed to working with his hands as a path to respecting the nature of the materials. He is also an avid violinist and DIYer.

MATERIAL
Corrugated cardboard

MATERIAL PROPERTIES
Simplification, Recyclable, Recycled Content

In 1922, the Russian suprematist artist El Lissitzky wrote a children's tale, *About Two Squares: In Six Constructions*, in which a black square and a red square fly to earth from afar. Concerned as he was with geometric form, Lissitzky was enamored of the square. More than ninety years later, the straight-edged quadrilateral form still fascinates—with surprising results.

For two engineering students at the Cooper Union, New York, their interest in right angles, rapid prototyping, and a shipping carton that could be opened without the use of a tool combined to produce, literally, a better cardboard box. Accustomed to working in an environment where trial and error, keen observation, and comparative evaluations are the dominant skills, the two drew up a sketch on paper, and then quickly moved to a 2D model in AutoCAD. Working with a laser cutter, they produced multiple versions of a flat unfolded box, which was then assembled, tested, and revised again and again until the box could be rough-handled and opened with only a pair of hands.

In the final design, "there are a lot of little details that go unnoticed unless you examine the cutout very closely," the duo explain. Origami-like, a flat sheet of cardboard transforms into a sculptural object. Free from the burden of who should or should not possess the skills necessary to design, the two are able to see the elemental in a whole new way.

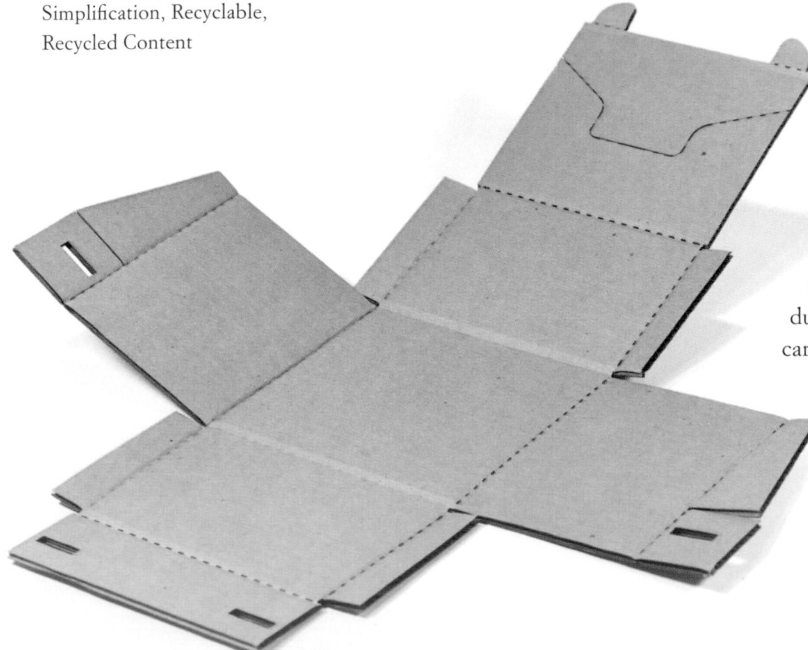

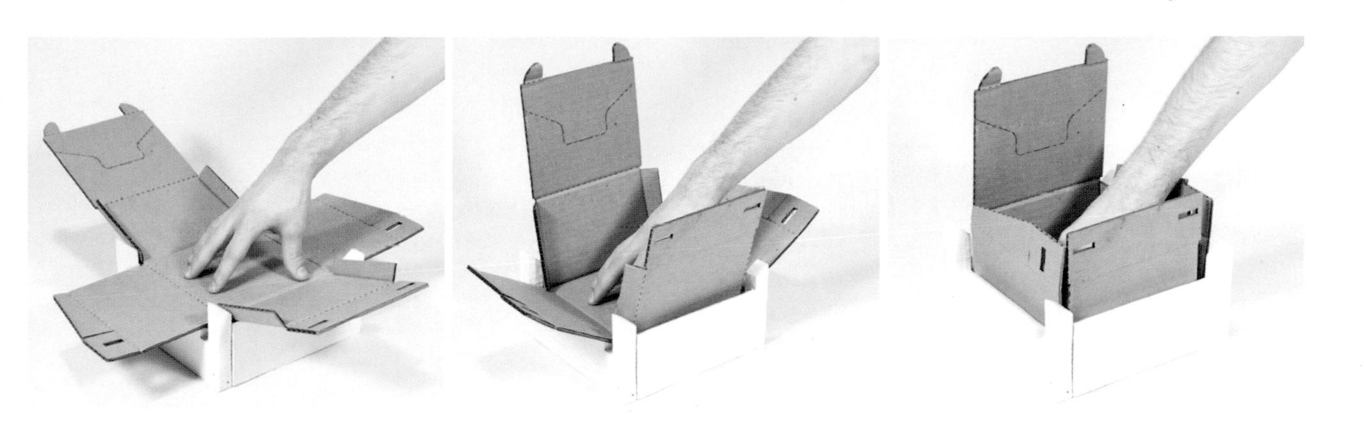

MATERIAL INSIGHT

Using only standard corrugated cardboard, this solution to creating a box from a flat sheet of material has significant advantages over more traditional methods, requiring 15 to 20 percent less cardboard and no glue, and markedly speeding up production and disassembly. Precision cutting from a laser or die cutter is needed for the overall outline, with a jig enabling quick formation into a box shape. Perforations are used to make construction easier, and are created slightly askew to aid the use of the jig. The construction also means that the box can be turned inside out and still be closed and opened, giving it a second life without shipping labels on its exterior. Though used for corrugated cardboard, the concept can be applied to other materials within the packaging and engineering space, with scalability of size a simple matter of changing the dimensions on the laser cutter.

opposite By designing the bottom of the box as a single, structurally sound piece of cardboard, 15–20 percent less material is needed than with traditional paperboard packaging, while maintaining a 118-kilo (260-lb) load strength.

this page To achieve the right shape, an unfolded Rapid Packing Container is pressed into a custom jig, then closed at the top and sealed with an adhesive strip.

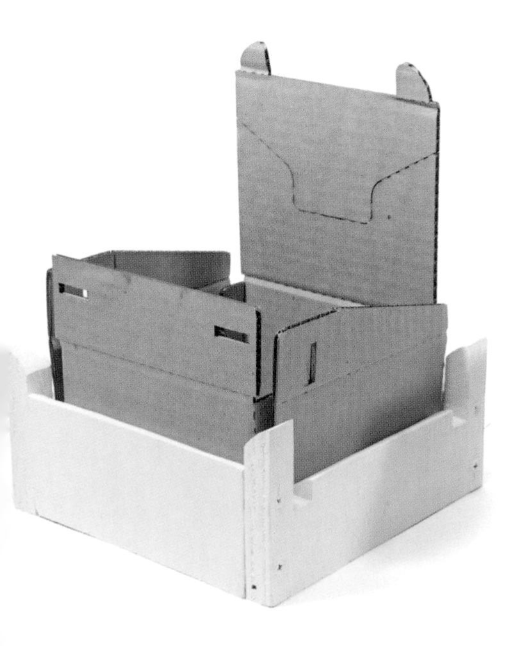

NATURAL 4X LAUNDRY DETERGENT BOTTLE

DESIGNERS
Seventh Generation and Ecologic Brands
www.seventhgeneration.com
www.ecologicbrands.com

DESIGNER BIOGRAPHIES
Peter Swaine is director of packaging at Seventh Generation. Julie Corbett is founder and CEO of Ecologic Brands, where she collaborates with senior industrial designer Rob Watters and industrial designer Romeo Graham to produce packaging solutions that eliminate the need for rigid plastic and use limited resources more efficiently.

MANUFACTURER
Ecologic Brands
www.ecologicbrands.com

CLIENT
Seventh Generation

MATERIAL
Pulped-paper fiber, low-density polyethylene (LDPE) and high-density polyethylene (HDPE)

MATERIAL PROPERTIES
Recycled, Recyclable, Reduced Material

By pressing on the side seam, the fiber shell pops open, allowing the user to compost or recycle the exterior with paper and cardboard, while the cap and liner can be recycled separately with plastics.

It might seem unlikely for a bottle of laundry detergent to earn a place in the lexicon of innovation. How much difference, really, can a soap container actually make? But with the advent of global warming, and with less environmentally responsible nations viewing the melting of the polar ice caps as merely an invitation to drill deeper in fresh territory, every small effort counts.

If you could attach a theme to Seventh Generation's Natural 4X Laundry Detergent container, it might be "new uses for old papers." The pulpy yet stiff exterior of the chunky detergent package has the tactility of an earlier era combined with the lightweight organic aspirations of the future. Impressively, all the packaging components (fiber bottle, pouch, spout, cap) use 66 percent less plastic than a typical 3-litre (100-fl. oz.) laundry bottle.

"Whereas most packaging companies innovate on the equipment side, our focus was always the consumer, to offer this audience a choice beyond plastic or glass. The ability to choose a sustainable package made from cardboard is entirely new," explains Seventh Generation. Even if consumers intuitively understand that beneath the surface is a plastic pouch filled with liquid suds, the exterior prompts them to think twice about the influence packaging has on the environment, if not their purchasing decisions.

Encouraging widespread change is not easy. But big brands can have a big influence. Well-marketed and -designed packages offer valuable models for how to adapt new materials, methods, and devices toward reducing our dependence on finite resources and reducing landfill.

MATERIAL INSIGHT

The clearly "sustainable" visuals of this packaging developed by Ecologic belie the level of technological development behind this hybrid molded pulped-paper container. The formed shells of the outer packaging are hinged, so it can

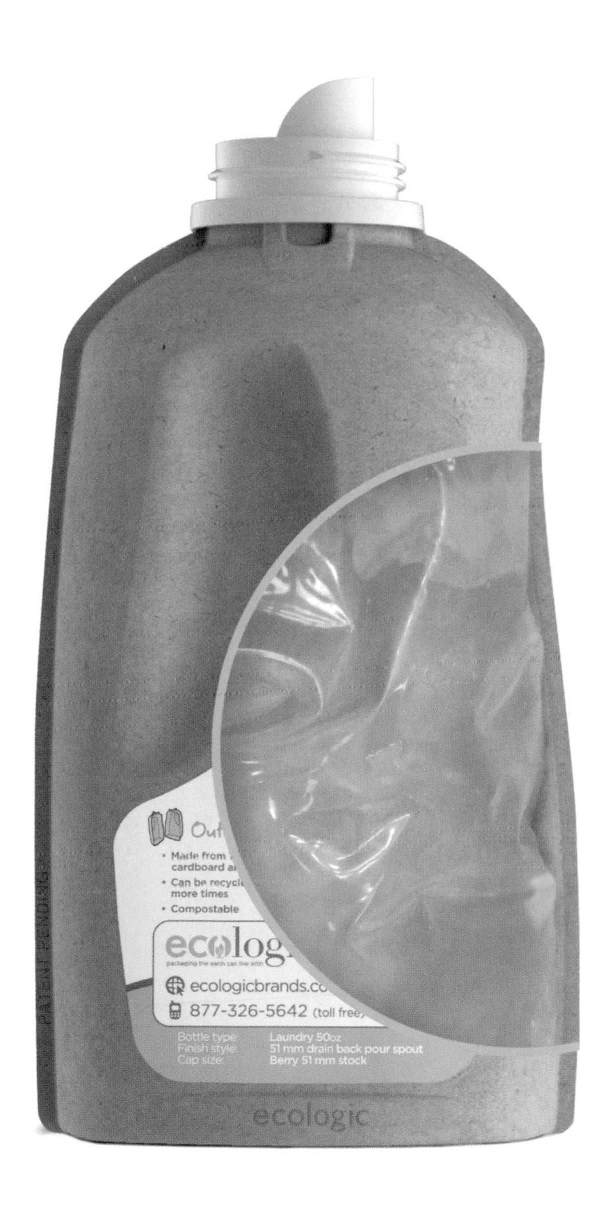

be shipped flatter prior to assembly around the flexible low-density polyethylene (LDPE) bag that will contain the detergent. The shell is then bonded around the bag and neck of the spout before filling. Though it is technically possible to produce paper cups and spouts with a molded plastic interior that can withstand moisture, the use of a more recognizable material (high-density polyethylene, HDPE) for this "touch point" gives greater confidence in the pouring and storing performance of the packaging. In addition, it is compatible with standard fill lines when adding the detergent. The outer pulped-paper container, as well as the inner LDPE bag, can be pulled apart after use for recycling.

Materials can talk. For Seventh Generation, a paper rather than plastic exterior conveys the company's aim to restore the environment and inspire conscious consumption.

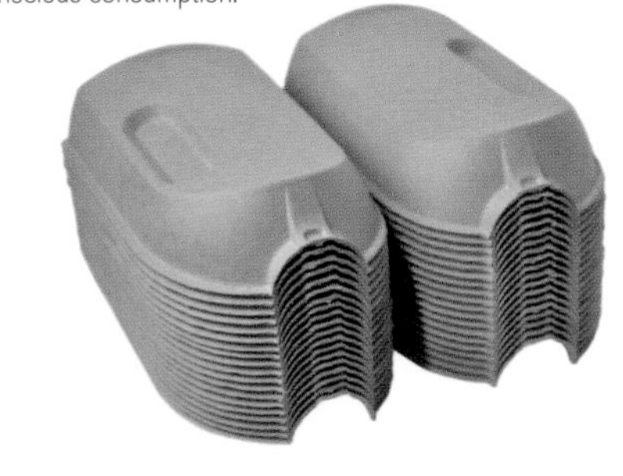

The molded fiber shell reduces the packaging footprint by using 66 percent less plastic than a comparable laundry detergent bottle.

THE CUBE

DESIGNER
Luis Felipe Rego

DESIGNER BIOGRAPHY
Luis Felipe Rego is the inventor of CUBE® and the director of research and development at Smart Packaging Systems, an affiliate of Compadre. Smart Packaging Systems was founded in 1992 to develop packaging solutions that combine raw materials and advanced technology for the paperboard industry. Winner of the DuPont Packaging Award for Innovation and Sustainability, it also provides graphic and industrial design services.

CLIENT
Various

MANUFACTURER
Smart Packaging Systems

MATERIAL
Paperboard and acrylonitrile butadiene styrene (ABS)

MATERIAL PROPERTIES
Lightweight, Compostable, Biodegradable

For much of the twentieth century, big corrugated-paper companies controlled the packaging industry. The research and development team at Smart Packaging Systems invented the Cube, a self-sustaining packaging system made from a specially engineered 100 percent paperboard lamination held together with super-strong, humidity-resistant glue. The design functions well and looks good all the way from the manufacturer production line to the delivery truck and out onto the retail sales floor. Based on the concept of open architecture crating, the Cube disrupted long-established industry practices while fulfilling strict environmental directives from such large-scale retailers as Walmart.

Far more than mere box-makers, the team then sought out new challenges, using modular systems as their primary mode of expression. In a departure from the usual practice of separating crating and display into different functions, they developed the Cube Display and the Cube Pak, reusable and economical alternatives to corrugated cartons and wooden crates that can be knocked down for flat-pack storage yet easily assembled without the use of tools. The designs straddle the divide between temporary displays, which can be flimsy, and permanent displays, which can be an investment, by offering an "extended life display" that lasts up to two years. "Previous solutions either haven't been enough or have been too much. This is something new in an industry that historically has not seen much change."

The Cube's systematized, Lego-like approach surpasses traditional point-of-purchase displays by conforming to standard shipping dimensions while allowing for customization by individual brands.

MATERIAL INSIGHT

This simple yet effective concept is made possible through the performance of the materials, specifically using paperboard as an engineering material. The edge pieces are produced from basic fiber-based paperboard, but bonded using a high-strength binder that is also humidity-resistant. The layers of paper are cross-laid, increasing the composite strength in all directions. The original designs for this innovation, known as the Cube, also used paperboard corner pieces, but this update, called the Cube Pak, instead uses injection-molded acrylonitrile butadiene styrene (ABS). These pieces are bolted to the paperboard edges to give even greater strength, and are sufficiently strong to be used as a point-of-purchase display, made more visible if the pieces are printed onto. Simple sheets of corrugated cardboard are used to fill in the open sides of the box. The overall structure can be easily broken down for either recycling or reuse.

above More than fancy gadgetry, retailers want display systems with the flexibility to conform to the space requirements of individual stores.

below Stackable components have been engineered for quick and easy assembly and disassembly to cut down on labor and encourage reuse.

THE AIDPOD KIT YAMOYO

DESIGNER
Pi Global Structural Team
www.piglobal.com

DESIGNER BIOGRAPHY
The Pi Global team has more than thirty years' experience in structural design and engineering for consumer goods and healthcare brands. The team includes production director Eric Connolly, senior development engineer Dave Salmon, and senior structural designer Jay Hussain. Their goal is to challenge the production status quo in order to deliver innovative packaging solutions for clients. Their effective, brand-relevant structural design and engineering solutions respect production schedules.

CLIENT
Colalife
www.colalife.org

MANUFACTURER
Charpak, Amcor Flexibles, Packaging Automation Limited

MATERIAL
Recycled polyester (rPET) and P-Plus®

MATERIAL PROPERTIES
Lightweight, Second-Life, Low Cost

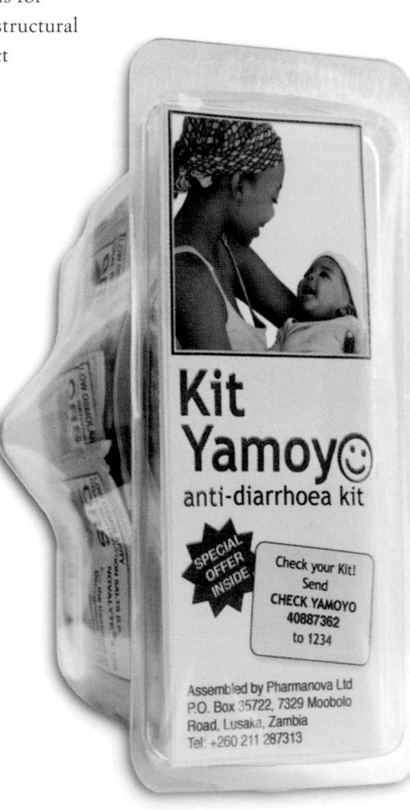

right The Kit Yamoyo has a flat top and undulating bottom to conform to the negative space in Coca-Cola packing crates, the anti-diarrhea kit's delivery system.

opposite The idea of an aid worker who observed that Coca-Cola is available in remote African villages yet medicine to treat diarrhea, the second biggest killer of children under five, is not, the package helps illiterate parents measure doses.

In its long history, Coca-Cola has been both celebrated and denigrated for creating one of the world's most recognizable brands while selling soda to developing countries where many children die without basic healthcare. The drink's famous contoured bottle is as instantly recognizable as the soda's frothy, exuberant taste, a sensuous yet impractical shape in an era concerned with efficient packaging and minimal waste.

So when Simon Berry, an aid worker in Zambia and co-founder of the charity Colalife with his wife Jane, sought a solution for how to reduce the number of people dying from diarrhea by getting ORS (oral rehydration salts) into their hands, the bottle's negative space sprang to mind. They devised the AidPod, a solution that piggybacks onto the soft drink's existing global supply chain by filling the unused space around the curvy glass package with lifesaving products, thereby establishing a delivery system to remote African communities.

The AidPod's wedge-shaped design enables packets of medicine to fit into the voids between the necks of Coca-Cola bottles. Inside the AidPod is a Kit Yamoyo ("Kit of Life") anti-diarrhea kit. It includes an instruction leaflet, zinc tablets, ORS, and soap; the pack itself acts as a measuring device for the water needed for a serving of ORS solution, a storage container, and a drinking cup. Ultimately, the pack can be further adapted to supply other life-saving medical kits.

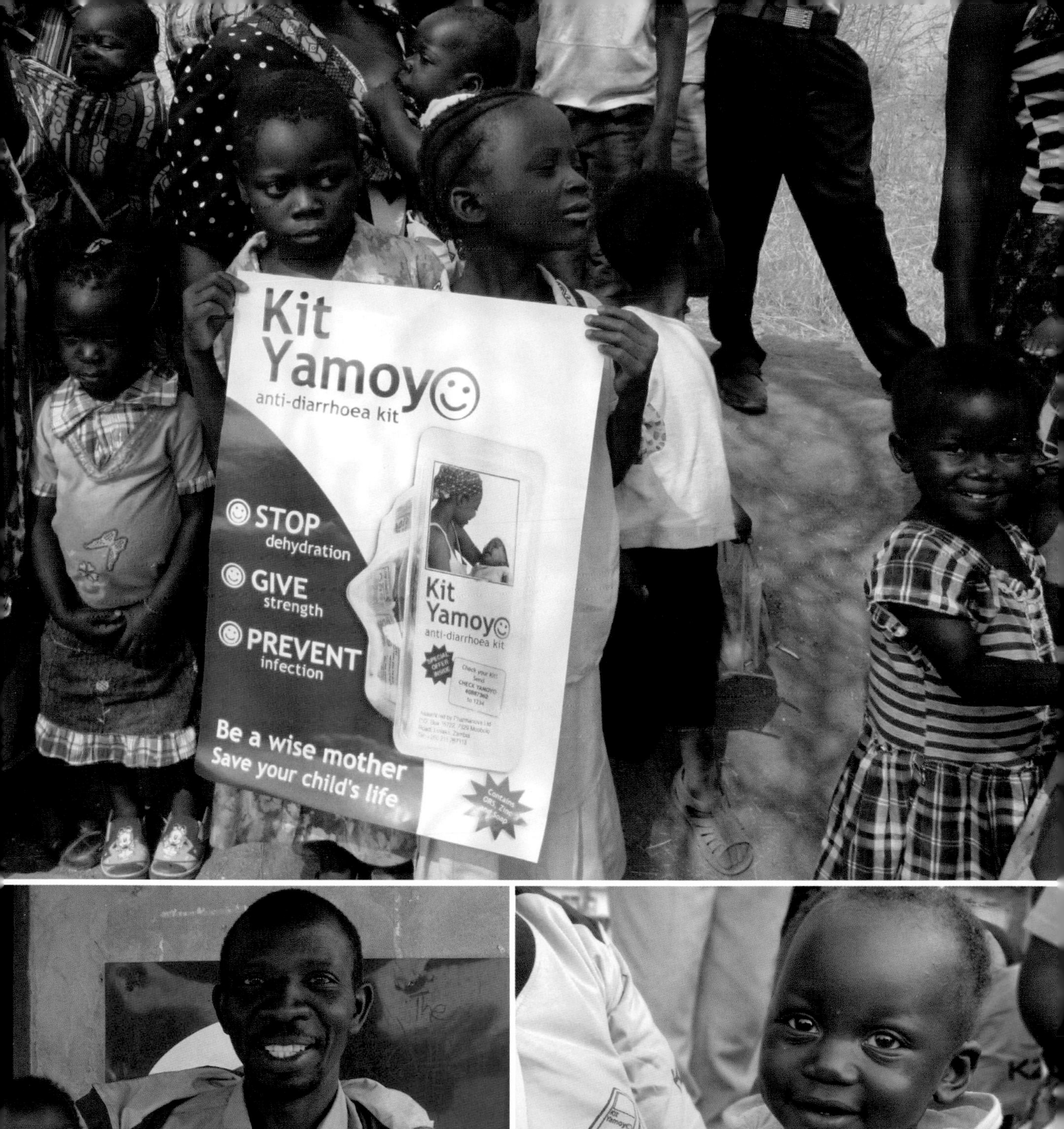
Kit
Yamoyo
anti-diarrhoea kit
STOP
dehydration
GIVE
strength
PREVENT
infection
Kit
Yamoyo
anti-diarrhoea kit
Be a wise mother
Save your child's life

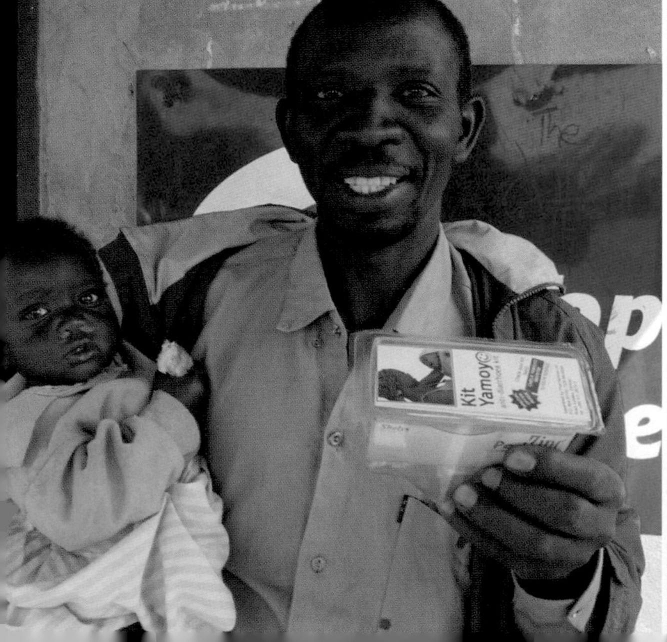
Kit
Yamoyo

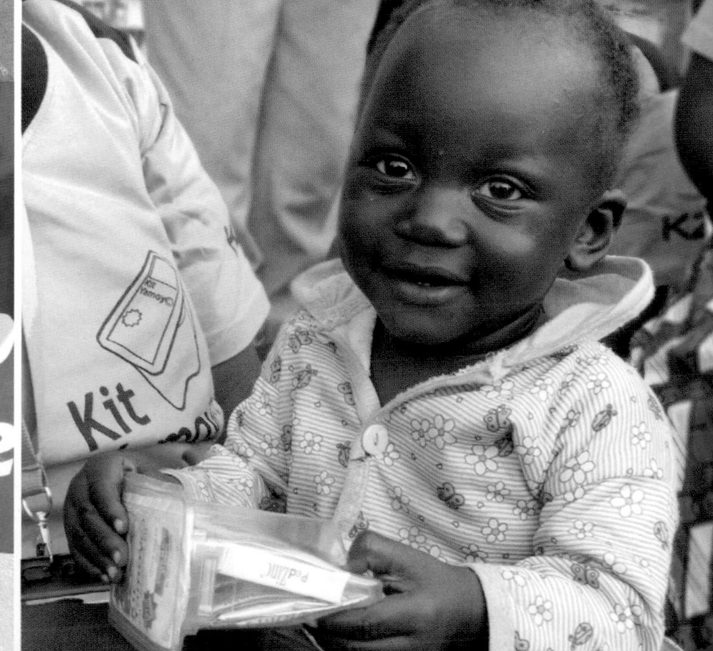
Kit
Yamoyo
Kit

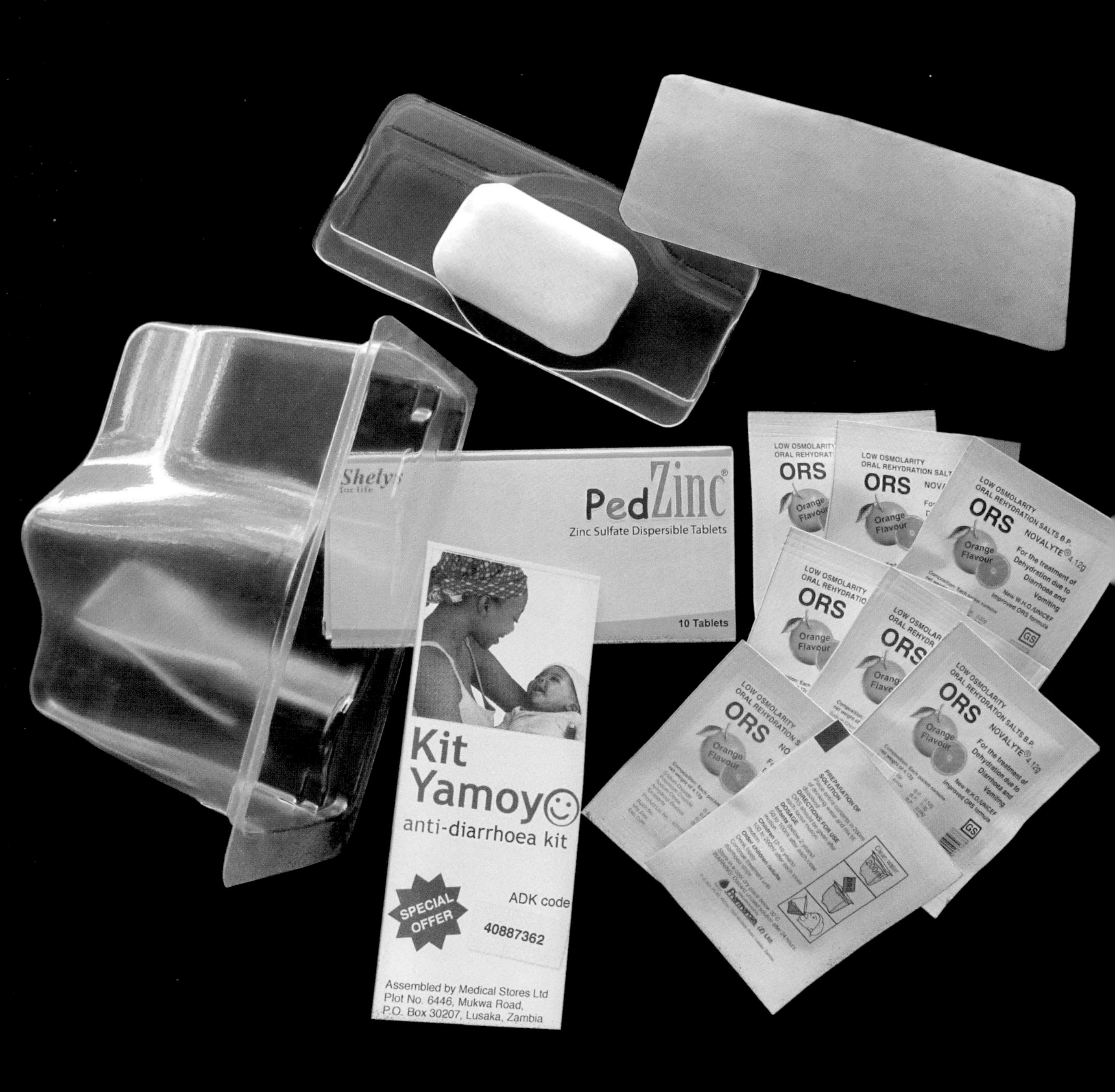

PedZinc
Zinc Sulfate Dispersible Tablets
10 Tablets
Kit
Yamoy☺
anti-diarrhoea kit
SPECIAL OFFER
ADK code
40887362
Assembled by Medical Stores Ltd
Plot No. 6446, Mukwa Road,
P.O. Box 30207, Lusaka, Zambia
LOW OSMOLARITY
ORAL REHYDRATION SALTS B.P.
ORS
NOVALYTE® 4.12g
Orange Flavour
For the treatment of Dehydration due to Diarrhoea and Vomiting
New W.H.O./UNICEF Improved ORS formula

opposite The container was designed from the inside out in order to accommodate sachets of Oral Rehydration Salts (ORS), zinc, soap, and an instruction leaflet.

right The packaging acts as the measuring device, a mixing and storage device, and a drinking cup.

below What started with maximizing the empty space in a Coca-Cola shipping crate as a vehicle for distribution has come to exemplify the role good design can play in saving lives.

MATERIAL INSIGHT

The molded transparent polymer outer packaging for the anti-diarrhea kit has a fourfold function: as protection for the contents; as a measuring cup, since many users do not have or understand measures; as a cup from which to drink the medicine; and as secondary storage for unused zinc and ORS sachets after the first dose has been taken. The molded container is produced from recycled polyester (rPET)—with 80 percent recycled content, mostly from beverage bottles. Its shape fits snugly between the returnable glass Coke bottles within a standard molded plastic crate. Additionally the reusable lid separates the soap from the other components of the kit and is itself covered by P-Plus®, a peelable, waterproof, microporous polymer film heat-seal top that is tamper evidencing. The material choices mean that the pack can withstand extreme conditions during distribution, including rapid temperature changes, rough transport, and altitude changes, reducing the risk of bursting while maintaining a hygienically sealed pack.

TETRA TOP

DESIGNER/MANUFACTURER
Tetra Pak
www.tetrapak.com

DESIGNER BIOGRAPHY
Tetra Pak is a leading food-processing and carton-packaging solutions provider. It works closely with customers and suppliers to provide safe, innovative, and environmentally sound products in more than 170 countries around the world.

CLIENT
Various

MATERIAL
Paperboard and high-density polyethylene (HDPE)

MATERIAL PROPERTIES
Recyclable, Aesthetic and Tactile, Simplification

Packaging designers recognize that one of their most important roles today is to help tackle the environmental crisis by showing consumers a path toward more sustainable living. This is a daunting challenge for anyone, given how deeply invested we are in the "better than the previous version" materialism that makes our global economy go round. It demands an acknowledgment that most of us have far more than we want or need, and that we are as responsible for making wise acquisitions as we are for a product's disposal.

Overwhelming though this may be, it is also filled with possibilities. At least, that is the approach Tetra Pak has taken. Viewing itself as a facilitator of the changes taking place around us, it looks for better ways to package and dispose of what we consume. Responding to consumers' increasing demands for improved recycling, Tetra Pak launched Tetra Top with Separable Top, a plastic top that can detach from the carton sleeve and be recycled separately, since mixing materials is an obstacle to efficient sorting. Even better, the option of the separable top is available to customers at no additional cost.

Tetra Top's main contribution to contemporary packaging has been to combine the convenience of a bottle with the style and feel of a carton. It is an ethos that avoids unnecessary complexity while driving recyclability as an element of design innovation.

MATERIAL INSIGHT

The increased popularity of hybrid packaging containers—those that use multiple different materials and constructions to enhance the performance of the container—has caused

What began in 1943 as a quest to improve the packaging of food products, in particular milk, during transportation evolved into a business culture of consistent innovation and a long-term commitment to sustainable practices at every step in the product's life cycle.

challenges with recycling at end of life. The Tetra Top with Separable Top container solves this problem for the packaging of chilled products such as milk by enabling the user to easily separate the two different materials: The top is made of high-density polyethylene (HDPE) for easy opening, pouring, and closing, while the body of the package is made of coated FSC-certified paperboard. The paperboard is perforated so that it is strong enough for normal use, but can be ripped apart when empty by simply pushing a tab into the container and pulling.

The ability to detach the plastic top from the carton sleeve responds to consumers' desire for packaging that facilitates recycling. A pre-cut perforation on the outer layer of the cardboard facilitates the separation of the top from the sleeve, without impeding the functionality of a design that seeks to deliver the benefits of a bottle and a carton in one.

REUSABLE SPOOL PACKAGING

DESIGNER
Satish Gokhale
www.designdirections.net

DESIGNER BIOGRAPHY
An alumnus of the National Institute of Design, Ahmedabad, Satish Gokhale brings more than twenty-five years of experience to his role as head of product design at Design Directions. He is a past member of the India Design Council and a founding member of the Association of Designers of India (ADI).

MANUFACTURERS
Consta Cool Pvt Ltd and Ideal Industries
www.consta.in
www.idealindia.com

CLIENT
Lakshmi Card Clothing Manufacturing Company Pvt Ltd
www.lakshmicardclothing.com

MATERIAL
Linear low-density polyethylene (LLDPE)

MATERIAL PROPERTIES
Lightweight, Impenetrable, Ergonomic.

Design Directions' reusable spool packaging speaks to the profound and enduring beauty of simplicity and clarity, even when applied in an industrial context.

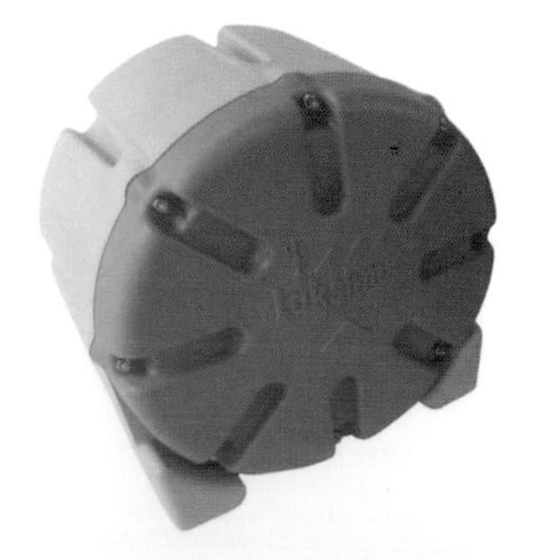

There is one aspect of the reusable spool packaging that speaks to the human-centered promise of good design precisely because it seems so obvious: A set of radial indentations is placed along the curved edge of a large plastic drum, in the right place and of the right size for an adult hand to easily grasp and lift. This makes it possible for a worker to comfortably carry the bulky container without risk of injury to the contents or to him- or herself. Its innovation lies in its improved usability but also in its assurance of the proper upright orientation through a concentration on the essentials. Rather than an attached strap or handle, the notch is integral to the basic form, making it unobtrusive and intuitively understandable.

It is the kind of improvement that might easily have been overlooked when considering the redesign of a package for a giant spool used to transport thin, flexible metal blades. Previously, the blades became entangled when the drum was transported horizontally; they were wound around spools that were first packed in bags to protect against moisture, then slid into metal drums and clamped with a circular ring. The drums were then set in wooden crates for transport.

The improved reusable version overcomes the problem thanks to its very obvious vertical form, meaning that it gets transported upright—it wobbles if transported horizontally. It has molded feet for proper orientation, and an ergonomic, self-explanatory approach that, when rendered in orange, delivers a jolt of color to the most utilitarian space.

MATERIAL INSIGHT

This attractive and effective update to traditional steel barrels for transporting heavy spools of steel blades uses rotational molding of linear low-density polyethylene (LLDPE). This molding process can easily create large-sized parts that are hollow (full-sized sofas can be molded this way); the rotation of the mold when it contains hot, molten plastic forces the resin only onto the outside of the cavity through centrifugal force, where it solidifies and creates the part. It allows for easy production of relatively complex shapes, as in this packaging, which has been designed to allow for easy hand-carrying and storage. The container seals when closed to minimize moisture ingress, which would rust the parts, removing the need for the metallized polyester foil bags necessary with the steel barrels.

above left four images In the previous system, to prevent damage during transport to the small, sharp teeth on the sides of the metal blades, each spool had to be carefully packed in layers by hand.

above No detail is too small, especially when it offers an intuitive guide to an ergonomic function—how to properly hold and lift an oversized object whose contents can be damaged if mishandled.

below A key feature in the conversion from the old to the new spool packaging is the eight grooves set along the rim of the lid and the base that pinch the spool, securing it against any internal movement.

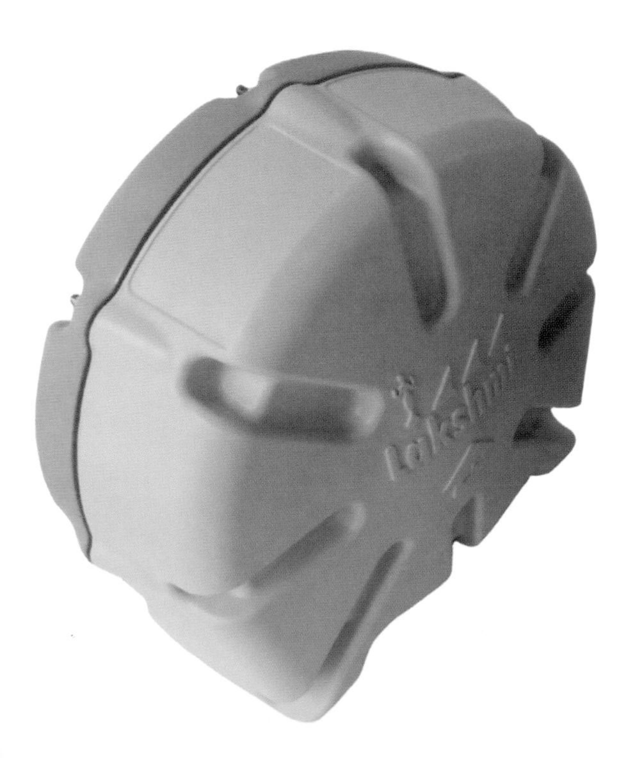

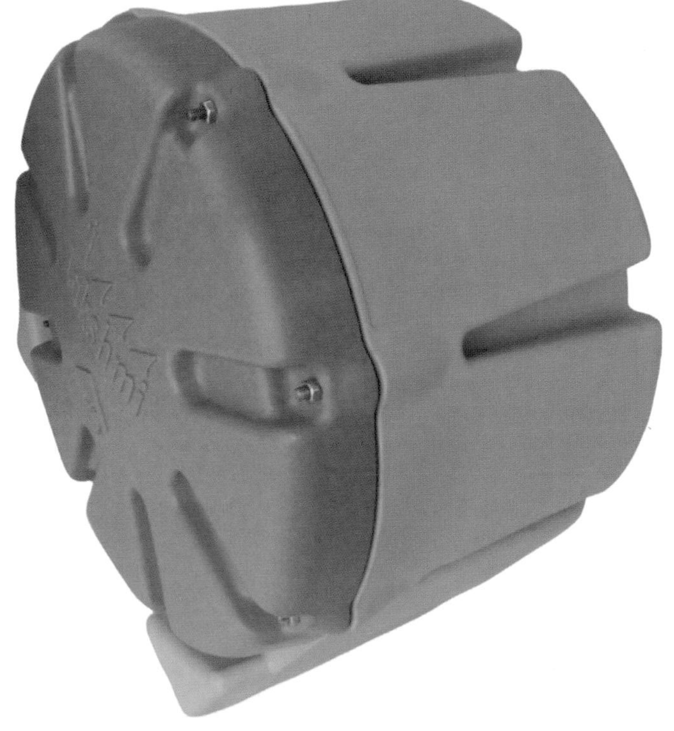

LEITZ ICON PACKAGING

DESIGNERS
Chris Whittall, Ari Turgel, Zack Stephanchick, Matt Taylor, and Dan Harden of Whipsaw Inc.
www.whipsaw.com

DESIGNER BIOGRAPHY
Whipsaw is an industrial design and engineering consulting firm located in Silicon Valley, California. Founded in 1999 by Dan Harden and Bob Riccomini, active principals in the firm, Whipsaw designs for diverse industries including housewares, consumer electronics, medical, and business products.

CLIENT
Esselte Corporation

MANUFACTURER
Leitz
www.leitz.com

MATERIAL
Recycled-paper pulp and recycled plastic

MATERIAL PROPERTIES
Smart, Recyclable, Tactile

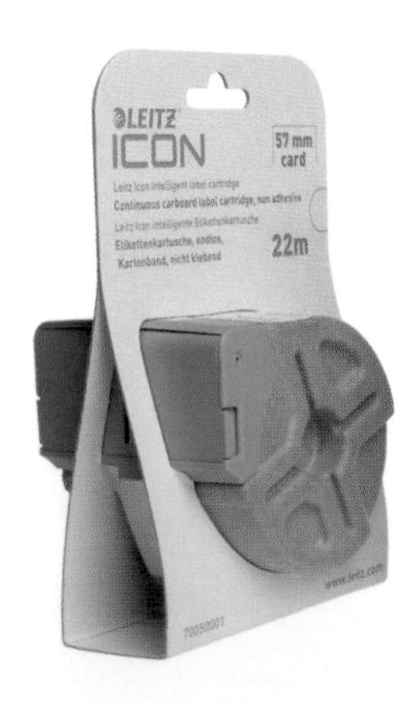

Recycled thermoformed paper pulp serves as both the cartridge body and the protective shipping box to achieve an environmentally responsible solution.

An unassuming printer cartridge at first glance, the Leitz Icon Smart Label cartridge is anything but. It is the guts of a portable, wireless labeling system so perfectly restrained and understated in its design that it leaves plenty of room for its primary tasks: identification and even self-expression. Unburdened by non-essentials, the design is consistent with the pared-down aesthetic of its creator Whipsaw, a Californian-based design and engineering firm that has built a reputation for simple elegance amid the competitive techno-crazed arena of Silicon Valley. While the firm has garnered awards for a breast-shaped baby bottle, a portable modem that creates a hot-spot on the go, and the game-changing Nike+ Fuelband, it also tackles less eye-catching challenges in areas that are ripe for innovation.

Before the Icon, label printers were unwieldy, difficult to load, and low-tech. So Whipsaw started from scratch. First they redesigned the label printer, giving it wireless capabilities so that it can print from any computer, tablet, or smartphone. Then they turned their attention to the label roll itself, cradling it in a cartridge made of recycled thermo-formed paper pulp. The fibrous material offers a tactile counterpoint to the printer's sleek contours, serving a dual purpose as a shell for the cartridge and as a shipping container. The green recyclable visor houses a microchip that conveys the roll size to the printer and the available label quantity to the user.

MATERIAL INSIGHT

Molded waste-paper pulp proved to be the ideal material solution for this adhesive-label printer cartridge. It is lightweight, easily molded and recyclable, contains recycled content, and provides structural protection for the labels inside. Because the cartridge functions as both the cartridge body and its own protective shipping box, a small recyclable

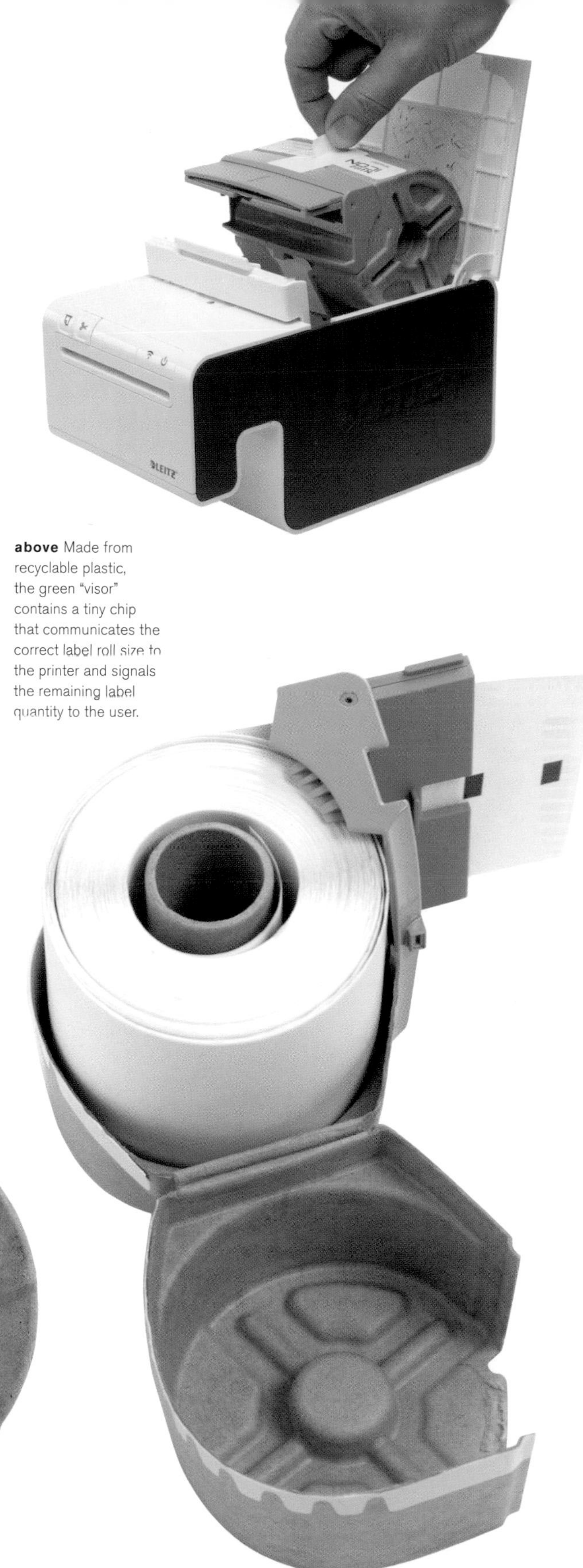

polypropylene (PP) part was integrated where the paper exits the cartridge. The size of this part was reduced to a minimum and assigned an abundance of duties, including adding structural integrity and dimensional stability to the product. While the green plastic part accomplishes much, the paper pulp provides considerable structure to the product as well as adequate shock absorption. The paper-pulp body itself is constructed as one molded piece, which is folded along the center and then taped at the seams. The fold actually functions as the alignment feature for each cartridge when it is placed in the label maker.

above Made from recyclable plastic, the green "visor" contains a tiny chip that communicates the correct label roll size to the printer and signals the remaining label quantity to the user.

below and right Rather than look down on the mundane task of printing out labels, Whipsaw sought to elevate it, streamlining the entire experience through design.

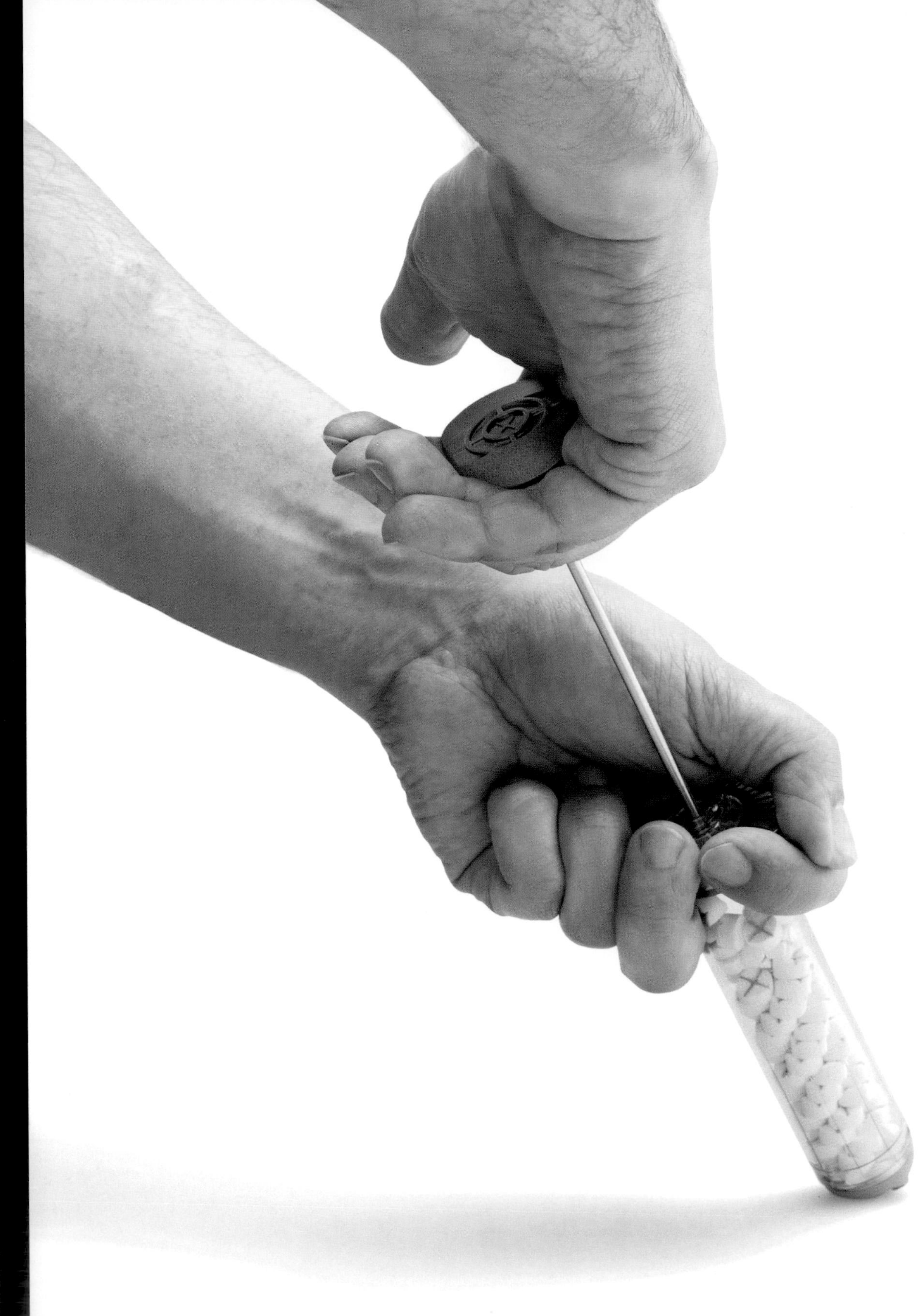

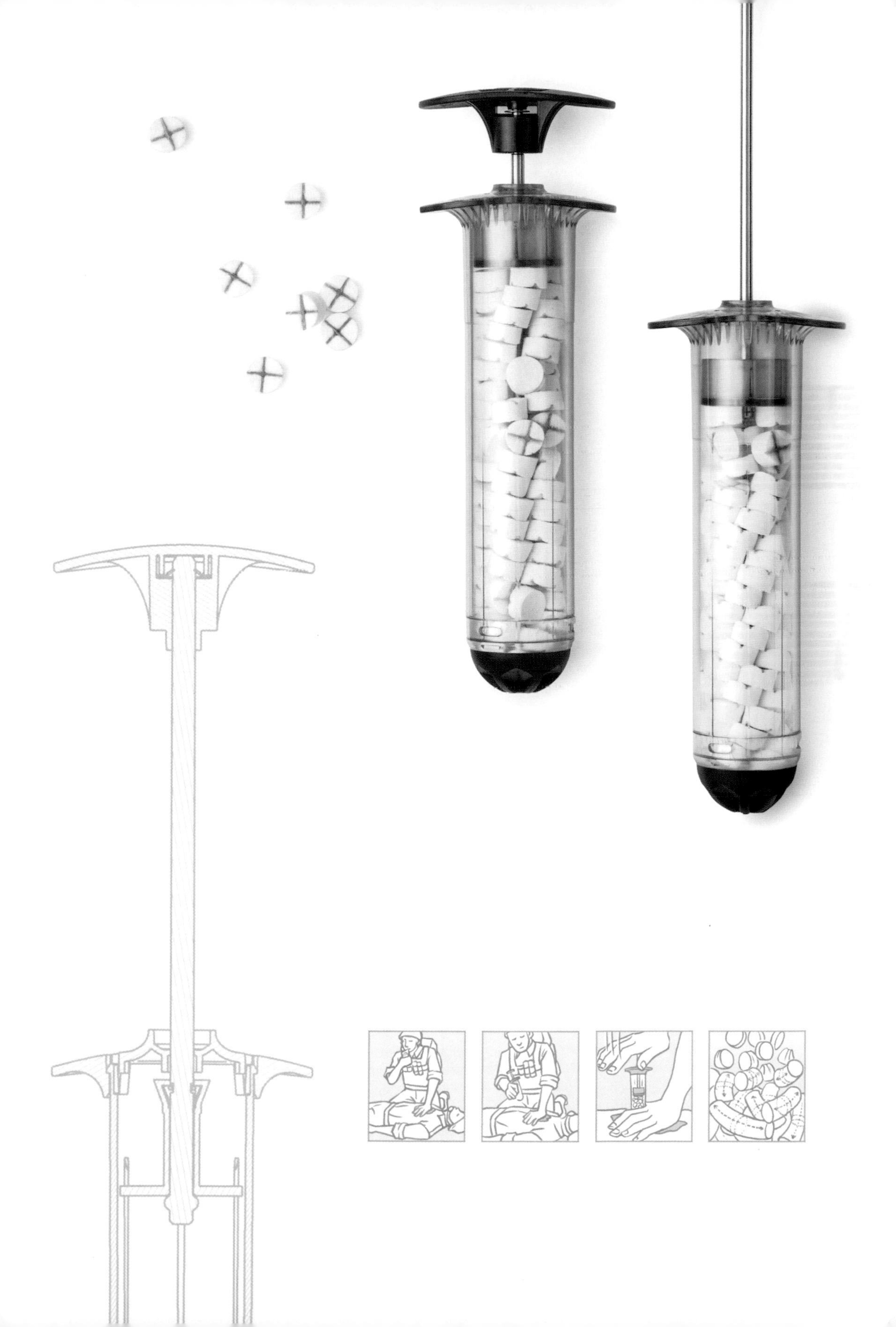

CHAPTER 3

DISPENSING SYSTEMS

This chapter covers "dispensing systems": packaging that enables enhanced or easier use or controlled and accurate dosage through the design and material selection for the package. As a result, these systems typically include the packaging material and also the dispenser, whether it is a separate machine or an additional part of or treatment to the existing package. Probably the least material-centric in this book, the innovations are more centered on design and mechanism, while utilizing materials as enablers rather than definers of the technology, as exemplified by the magnets used in the bases of the cups in the "Bottoms Up Draft Beer Dispensing System"; magnets are not new, but their incorporation enables beer cups to be filled at a rate unthinkable without the technology.

Dispensing systems are predominantly a result of providers' or consumers' desires for easier use and more controlled delivery. For example, consumers' search for the "perfect" cup of coffee every time has led to a category-killing upsurge in the use of pod dispensers such as Nespresso and Keurig that combine a high-performance machine with individually packaged portions of coffee to ensure a consistent and mess-free experience. Pod coffee makers now outsell drip coffee machines in the United States and Western Europe, suggesting that we like the ease and perfection of a dispensing system. The "podification" of other consumables such as tea, hot chocolate, and soup confirm this trend. The challenge now becomes one of price (much of the pod market sells coffee for effectively $50 per pound) and waste, since every cup produces at least one not always easily recyclable pod.

opposite Natural Machine's Foodini "dispenses" ingredients in the same way a 3D printer prints products, depositing layers of material in controlled patterns to build up the 3D object or, in this case, a pizza, cookie, cake, or other food product.

In many ways, additive manufacturing—the umbrella term for 3D printing methods—is a dispensing system, creating products through the CNC positioning of powdered or semi-molten raw material along 2D "print" lines and repeating the process to build up a 3D part. However, beyond the engineers, hackers, and makers, the claimed inexorable rise in popularity of 3D printers for personal use will be achieved only when systems can offer a "no mess, no fuss" option, whether they are for toys, replacement parts, or food. The vast majority of consumers require simplicity and ease of use in their products, and will quickly tire of a machine that requires maintenance and clean-up and does not offer a perfect product every time.

Perhaps one of the most advanced examples of this type—and a shining example of no-mess-no-fuss delivery—is the Coke Freestyle machine, an "all-in-one" device that offers almost every single drink product that the company produces, with an ability to mix them so as to produce unlimited new flavors and combinations. The system is

The full-color Chef Jet Pro offers creative bakers the ability to print 25 x 35 x 46 centimeter (10 x 14 x 18 inch) forms in chocolate, sugar, or candy. The objects shown are playful yet edible sugar "cubes."

based on the inkjet printer, and uses powdered flavorings rather than the traditional concentrated syrups, increasing speed of delivery and reducing the potential for mess.

Dispensing systems do not always need to be complex, and some offer innovation through simplicity. Folia Sugarcane, the Drinkable Book, and Spacklit in this chapter solve existing challenges in very simple ways, the latter using natural human movement and dexterity as an integral part of its functionality. In this way, these projects show the full intention of systems—that they should make things easier for the user in a way that does not necessarily require learning a new skill. The development of the XStat Syringe is a good example of this. Various concepts were posited, and some were actually more efficient than the final design; however, in a battlefield situation where the user is likely to be under extreme stress or even in shock, a product with a motion that is easily recognized—the pushing of a plunger into a syringe—was deployed more effectively than unexpected designs simply because it required no thought to operate.

One project that is very much about the material is LiquiGlide, which is itself an enabler for potentially a vast number of new dispensing systems. This "treatment" to render surfaces super-slippery such that nothing sticks to them has applications far beyond packaging, but can ensure that dispensers get every last drop of product out with minimal effort, a challenge for condiments, motor oil, hand soaps, toothpaste, and numerous other more viscous liquids and pastes.

Ease of use, consistent delivery every time, minimal waste or clean-up, and the application of existing human movements and behaviors are the hallmarks of effective dispensing systems, with materials playing an important part.

Amazon's working prototype of its "Prime Air" service, a drone delivery system that Amazon claims will get packages to consumers within thirty minutes of order confirmation.

XSTAT SYRINGE

DESIGNER
Eric Park, creative director, Ziba Design
www.ziba.com

DESIGNER BIOGRAPHY
Eric Park leads strategic design and innovation programs at Ziba, integrating user research, design planning, and technology to help clients better engage consumers. His award-winning work with clients ranges from startups to Fortune 500 companies, including Nike, Philips Healthcare, Intel, and Procter & Gamble.

CLIENT
US Military

MANUFACTURER
RevMedx Inc.
www.revmedx.com

MATERIAL
Wood pulp coated with chitosan (an antimicrobial, blood-clotting material), silicone, and polycarbonate

MATERIAL PROPERTIES
Compact, Lightweight, Ergonomic

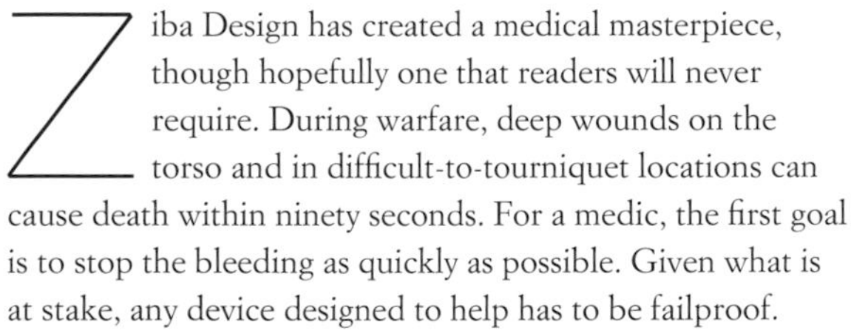

Ziba Design has created a medical masterpiece, though hopefully one that readers will never require. During warfare, deep wounds on the torso and in difficult-to-tourniquet locations can cause death within ninety seconds. For a medic, the first goal is to stop the bleeding as quickly as possible. Given what is at stake, any device designed to help has to be failproof.

So RevMedx, an Oregon-based startup focused on medical products, developed the XStat, a collection of small coagulant-saturated compressed sponges that, in a matter of seconds, decompress within the wound and expand outward to halt the bleeding without the traditional need for manual pressure or gauze. This frees up medics in the field to more rapidly attend to others, or to defend themselves if needed.

Ziba's task was to create a mechanism for administering the sponges, with stringent requirements: It had to be light and compact enough to fit inside a body-armor pouch, reliable at both cold and warm temperatures, and ergonomic for single-handed use. The solution is a syringe-like plunger with a perforated silicone membrane at the tip that secures the moisture-sensitive sponges until they have been inserted into the wound and ejected, filling the cavity and creating enough pressure to stop the bleeding.

Sleek and easy to use, the XStat does its job perfectly, and no more. When saving lives in seconds is the benchmark, that is enough.

MATERIAL INSIGHT

The syringe is manufactured from injection-molded polycarbonate (PC) that can withstand the rigors of a combat medic's bag, is highly transparent to ensure easy assessment of the contents, and is moldable into complex thin-wall parts for lightness and accuracy. The material has an impact resistance far greater

Along with compressed sponges that have been coated with a coagulant, the XStat syringe stores the plunger inside to reduce the overall size and make it easier to carry.

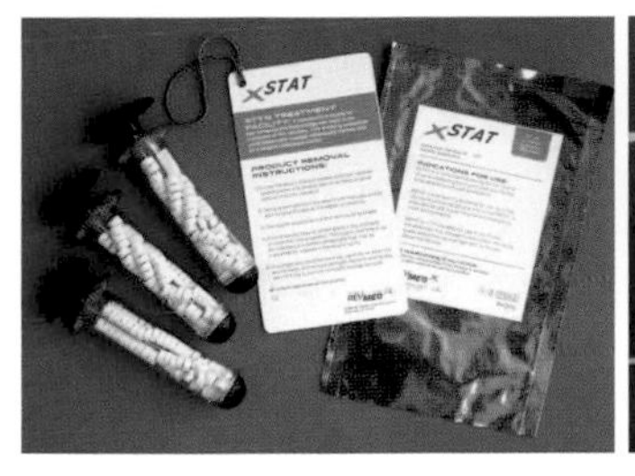

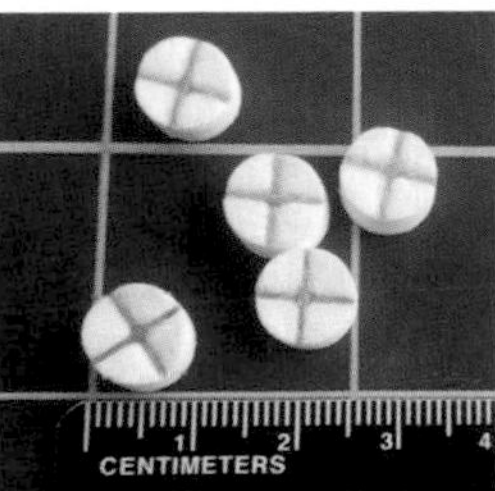

than virtually any other clear plastic. The syringe has an ambidextrous plunger that is designed to deploy the entire contents in one motion, which for most wounds is sufficient. For large-impact wounds, up to three applicators may be required. The sponges themselves expand instantly on contact with fluid, exerting enough pressure to stop arterial bleeding hands-free, and they are marked with blue strands that are visible via X-ray to ensure that all material is removed during surgery at the hospital.

left In a medic's field kit, bulky rolls of gauze are replaced with single-use XStat applicators. Each tiny sponge is marked with radiopaque threads.

below The lightweight polycarbonate syringe is designed for single-handed, ambidextrous use, and comes in different sizes to treat a variety of wounds.

FOLIA SUGARCANE

DESIGNER
Gabriel Collins

DESIGNER BIOGRAPHY
With an education that marries industrial design and business management, Gabriel Collins was a lead industrial designer at Eco-Products. Originally a family business founded in 1990 by father and son Kent and Steve Savage, today Boulder, Colorado-based Eco-Products is the leading manufacturer of sustainable single-use food-service products in the United States and Canada, a brand widely regarded as being at the forefront of innovation in this fast-growing sector.

MANUFACTURER
Eco-Products
www.ecoproducts.com

CLIENT
Various

MATERIAL
Sugarcane

MATERIAL PROPERTIES
Renewable, Compostable, Microwave- and Freezer-Safe

Boulder, Colorado, has always stood apart from other American cities for its history of innovation with an environmental edge. So it is almost to be expected that an elegant alternative to the staggering amount of expanded polystyrene- (EPS) foam trash that is accumulated daily from fast-food takeout would be born there. Eco-Products creates disposable necessities without the frills. From the company's early days onward, they have addressed the need for food-service packaging made from recycled materials. As designers and manufacturers, they have grown increasingly skilled at structural designs, culminating in their patented Folia line, a neat system for which the performance is in the details.

Traditional clamshell packages are made from non-renewable petroleum, while Folia is made from sugarcane, a compostable resource that does not have a future in landfill. It is freezer- and microwave-safe. The food containers have interior compartments for cutlery and condiments, as well as rounded corners that inhibit spills. Most ingenious of all, though, are the lids: Perforated hinges allow them to

Replacing standard petroleum-based plastic clamshells, Folia is made from sugarcane, a rapidly renewable resource that can be composted to divert waste from landfills.

be ripped off and discarded so it is easier to eat straight from the container without battling flaps that fold back on themselves. A modular design, Folia comes in six different sizes that fit inside each other, while tactile sugarcane prevents slippage on most food-preparation surfaces.

Perhaps best of all, Folia protects food while improving the entire experience, with a design that shows what is possible when beauty is as valued as utility.

MATERIAL INSIGHT

These pulp-fiber-based food delivery containers combine a low-environmental-impact material with an "all-in-one" design that significantly reduces the number of additional packaging pieces. This in turn makes recycling or composting a much simpler process. The cartons are molded from bagasse and a binder into rigid, lightweight forms using heat and pressure. Bagasse is the fibrous matter that remains after extracting the juices from sugarcane and is a rapidly renewable, compostable fiber that is typically incinerated or discarded. These fibers are chopped up, cleaned, bleached, and formed into mats. Once combined with a binder, the paper-like sheet is easily moldable, resistant to hot and cold temperatures (at least those experienced during food-heating, delivering, microwaving, and serving), and water-resistant.

above top Built into the stackable design are such carefully engineered details as cutlery nooks, sauce caddies, and spill-resistant corners.

above bottom Beyond the ease of filling and storing, the container works with hot and cold foods, in microwaves, and in freezers.

DRINKABLE BOOK

DESIGNERS
Senior designer: Brian Gartside, DDB New York; graphic designer: Aaron Stephenson; chemist: Dr. Teri Dankovich
www.briangartsi.de

DESIGNER BIOGRAPHY
Graphic designer and typographer Brian Gartside graduated from Virginia Commonwealth University with a BFA in Graphic Design. He worked with Michael Gericke at Pentagram before joining DDB New York, where he was hired by Menno Kluin and Juan Carlos Pagan.

MANUFACTURER
Printer: Jamie Mahoney, Bowe House Press; Binding and 3D: Peter Ksiezopolski, xweet; 3D Printing: Morpheus Prototypes

CLIENT
WATERisLIFE
www.waterislife.com

MATERIAL
Paper embedded with silver nanoparticles that are lethal for microbes and food-grade ink especially for letterpress printing

MATERIAL PROPERTIES
Highly Antibacterial, Printable, Non-Toxic

Both in content and in form, the Drinkable Book, an initiative undertaken by the non-profit organization WATERisLIFE, embodies a quest to change the lives of millions of children under the age of five who die each year from water-, sanitation-, and hygiene-related diseases. The book is both an operating manual and a toolkit that seeks to educate those most at risk while extending them a lifeline of drinkable water. Each page of the book is made of thick, sturdy paper that has been coated with silver nanoparticles that transform it into

Packaged as a graphically elegant, slipcovered book, a three-month supply of humble yet vital water filters is elevated to a compelling educational tool and a work of modern aesthetics.

a highly antibacterial water filter, offering greater meaning and usefulness than either a book or a filter alone could do.

The Drinkable Book is rooted in research work conducted by Teri Dankovich while a PhD student at McGill University, and later at the University of Virginia. Embedded with ions that ultimately have the potential to eradicate disease by halting transmission, the pumpkin-hued paper is packaged as a slipcased book that protects the pages and provides a frame for each filter, holding a page taut as water passes through and is purified. Developed for use during educational sessions on maintaining a clean water source in such places as Haiti, Africa, and India, each book is capable of providing someone with clean water for up to four years. Most importantly, explains Dr. Dankovich, "All of the chemicals used [to treat the paper] have been selected because they are safe, renewable, and non-toxic."

MATERIAL INSIGHT

Though perhaps not the most efficient or compact of water filters, it does the job. The paper is treated with silver nanoparticles that act as a broad-spectrum antimicrobial, purifying the water as it passes through the fibers of the paper sheet. Though initially white, the paper turns yellow during treatment, eventually becoming a deep golden color over time. Silver can safely be ingested in very small quantities, and, thanks to the very small particle size, it is used in tiny amounts while still maintaining complete efficacy in killing waterborne bacteria and viruses. The ink is printed by letterpress, the entire machinery cleaned to ensure zero contamination. The ink used is food-safe, the same used for printing on the inside of food packaging. The book packages multiple pages, each of which is two 29-centimeter-square (4.5-inch-square) filters. Each filter square can purify 100 liters (211 US pints) of water, or approximately one month's supply for an adult male.

To reduce the spread of diarrheal diseases, millimeter-thick pages containing silver nanoparticles filter out microbes, leaving behind water that is safe enough to drink.

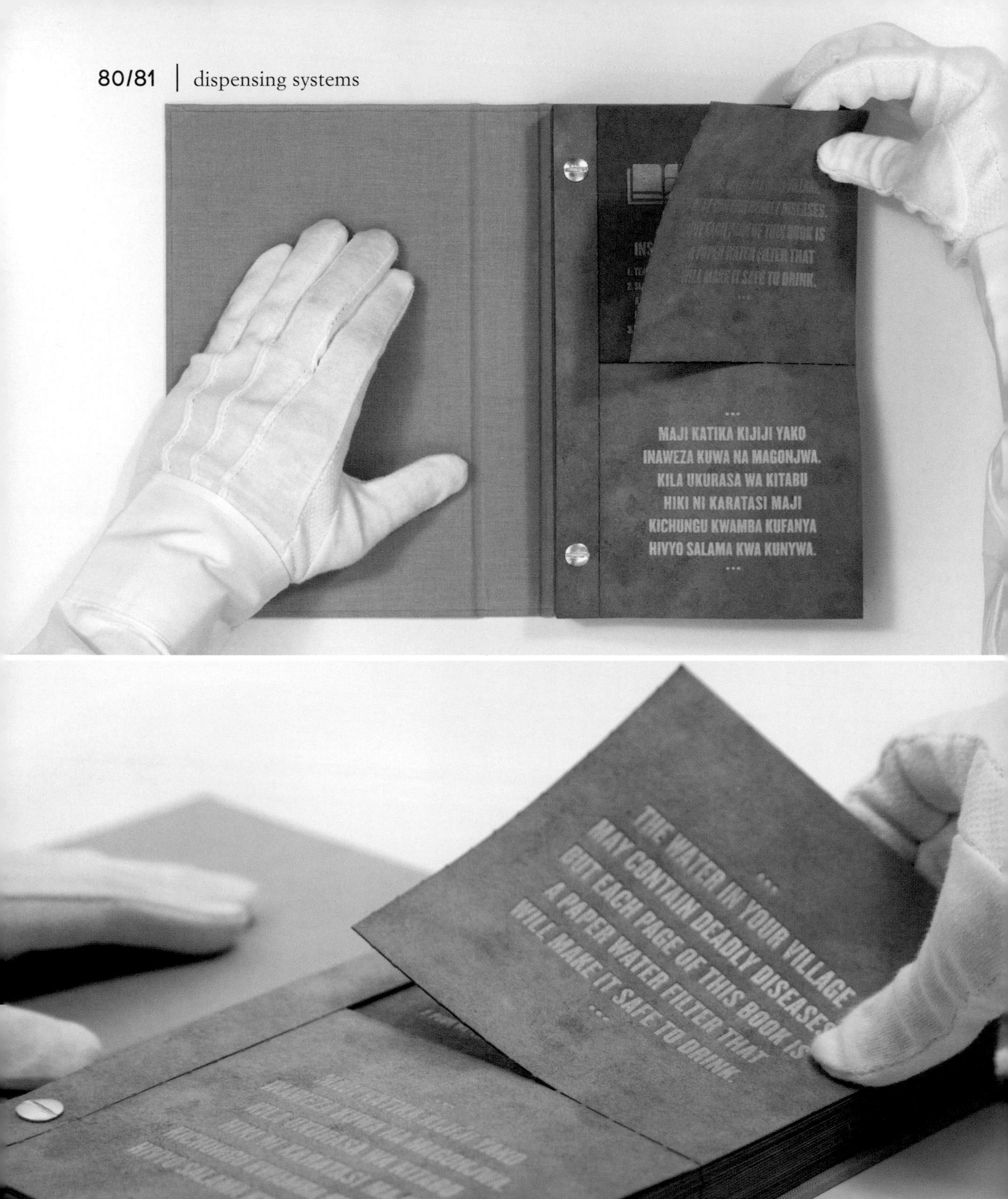
MAJI KATIKA KIJIJI YAKO
INAWEZA KUWA NA MAGONJWA.
KILA UKURASA WA KITABU
HIKI NI KARATASI MAJI
KICHUNGU KWAMBA KUFANYA
HIVYO SALAMA KWA KUNYWA.
THE WATER IN YOUR VILLAGE
MAY CONTAIN DEADLY DISEASES.
BUT EACH PAGE OF THIS BOOK IS
A PAPER WATER FILTER THAT
WILL MAKE IT SAFE TO DRINK.

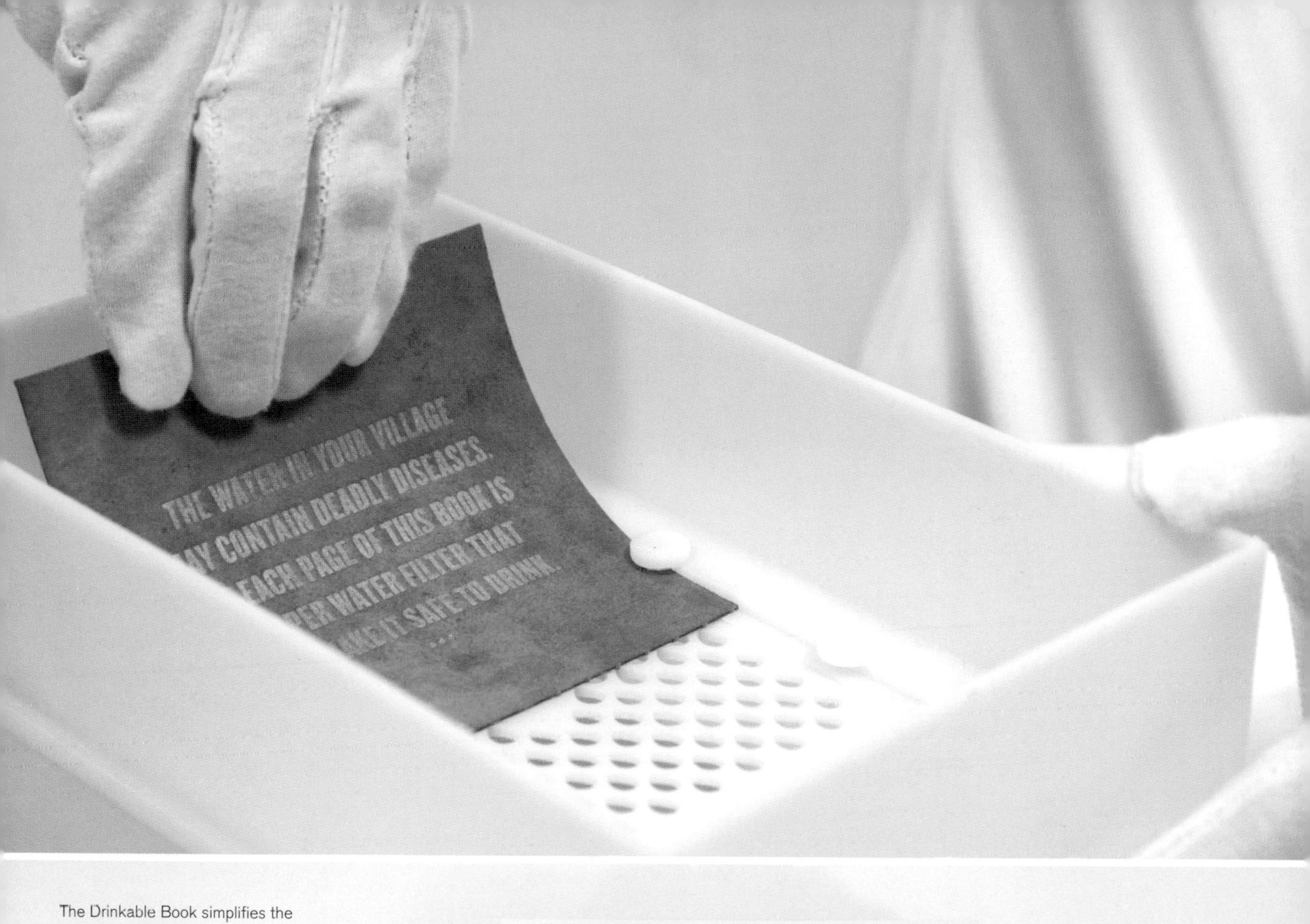

The Drinkable Book simplifies the daunting global water crisis into an easy-to-use manual and conveniently packaged tool that aims to save lives by providing the reader with a way to create clean, potable water from each page.

LULLABY PAINTS ECO-POUCHES

DESIGNER/MANUFACTURER
Imperial Paints
lullabypaints.com

DESIGNER BIOGRAPHY
South Carolina-based Lullaby Paints' baby-safe paints are made from the finest ingredients to meet the highest environmental standards of purity and safety without compromising exceptional color, coverage, durability, and finish. There are no fillers or solvents such as are found in traditional paints.

CLIENT
Various

MATERIAL
Multilayered polyolefin films

MATERIAL PROPERTIES
Malleable, Non-Toxic, Robust

It is no small claim to bill yourself as "the world's best organic paints." True or not, it gives Lullaby Paints a luxurious position accompanied by an expectation of absolute purity and superior pigment. A pioneer in the development of toxin-free paint that is sought after for baby nurseries, Lullaby comes, not surprisingly, in classically soothing and serene tones. Such colors as "starlight," "cotton candy," and "snowy fleece" show they are unafraid to be pretty and sweet, yet they escape the saccharine through advanced formulations and the use of pouches for their packaging.

Most of today's paints, finishes, and furnishings are treated with chemicals to help maintain them. Over time, these chemicals emit fumes known as volatile organic compounds (VOCs), whose accumulation has been linked to certain health risks (asthma, allergies, developmental disorders). So Lullaby created chemical-free paints that restore a sense of control and comfort to the pure at heart.

They took a similarly soft approach to their packaging. Metallic eco-pouches with cutout handles and spouts akin to a squishy watering-can introduce a language of malleable yet protected volumes. Pouches weigh 50 percent less than cans, which means less energy to manufacture and ship, plus more pouches per delivery since they better conform to the interior of a carton. And because pouches can be squeezed then resealed, there is less waste. It is an effective solution with savings all round.

Flexible stand-up pouches come with numerous advantages: The shape and graphics are easily customized and they are amenable to closures that encourage reuse, such as zip locks, spouts, and screw tops.

MATERIAL INSIGHT

Another product succumbs to pouchification, the trend for using pouches that has spread inexorably through the food, cleaning, and even motor-oil aisles of the store. The pouch offers significant advantages in terms of shipping weight, container form, reduction in size when unfilled, and now, in its recyclable versions, the ability to have a more useful end of life. The eco-pouch also has a distinct advantage over metal or even molded-plastic paint containers in that the air can be removed after partial use to ensure a better quality of product even if used again years later. It is constructed from multilayered polyolefin films that protect the paint from oxygen ingress and light degradation. The multi-seal format aids stability when in use or during storage, and the recyclable cardboard is tough enough to deal with the abuse typically experienced during transit and can be removed once the paint pouch is opened.

Lighter and smaller in volume than conventional paperboard, glass, or plastic, pouches have been readily adopted by the food and beverage industry, in particular since they are especially useful for fluid contents—including, as in this case, paint.

AIROPACK

DESIGNER/MANUFACTURER
Airopack
www.airopack.com

DESIGNER BIOGRAPHY
Dutch-based Airopack Technology Group is led by CEO Quint Kelders. Since the mid-1970s, the Kelders family businesses have been innovators in plastics molding and technical packaging for pharmaceutical, health and beauty, and food products. Central to their success has been the development of mechanical and pressure-controlled dispensing packaging technologies and systems.

MANUFACTURER
Airopack Technology Group
www.airopack.com

CLIENT
Various

MATERIAL
Blow-molded plastic pressurized dispenser that uses air instead of chemical propellants

MATERIAL PROPERTIES
Recyclable, Moldable, Non-Pollutant

The appeal of a leaner, greener future is undeniable. But advances often come with a hidden cost that undercuts their merits. Yet Airopack, maker of a pressurized dispensing system for fluids and other high-viscosity materials (perfect for health, beauty, pharmaceuticals and food), refused to be dissuaded from the possibility of eliminating environmentally toxic propellants in packaging. A Dutch-based family-run business, it made the development of a sustainable pressurized dispenser its mission. During seven years of research and development it held true to that pursuit, ultimately producing a patented pressure-control device that works with nothing more than fresh, clean air.

At first glance, other than an unusual helix-like mechanism, there appears to be little that is truly innovative. But look closely at the base and Airopack's award-winning design reveals itself. A "high-precision pressure regulator" combines with an "air-pressure reservoir for constant and controlled flow"; the dispenser can be held in any position—CEO Quint Kelders calls it the 360—and the product is dispensed with the same even consistency from start to finish.

The latest skincare and cosmetic formulations are infused with active botanical ingredients and Airopack's reliance on air instead of conventional hydrocarbon chemical propellants is especially well suited to enhancing a message of sustainability for brands and consumers that are inspired by and invested in the natural world.

Expertise in plastics-molding and pressure-controlled technology plus a commitment to product innovation have been key to the success of Airopack's patented designs.

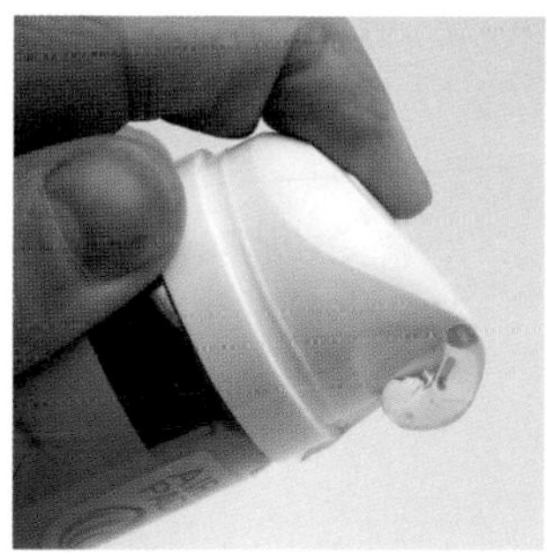

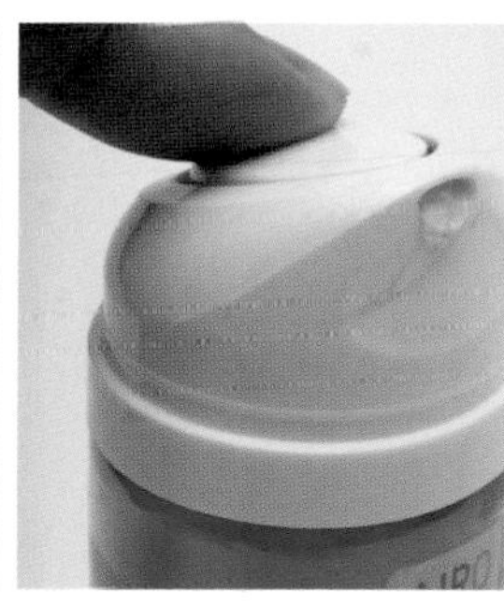

Airopack's unique technology for dispensing liquids, creams, gels, and foam offers an environmentally sustainable alternative to aerosols. No matter how the container is held, in any direction and even upside down, the product emerges with the same even consistency from start to finish.

MATERIAL INSIGHT

This patented pressure-control technology enables the dispensing of sprays, liquids, and gels with no hydrocarbon propellants. The system replaces traditional aerosol products used with conventional metal cans and chemical propellants. It is able to achieve a standard 2.0-bar pressure (for creams and gels) as well as the high 3.5-bar pressure necessary for sprays. The entire system is manufactured from polymer resins with no need for metal parts, making it fully recyclable. It uses compressed air rather than hydrocarbon chemical propellants, enabling the pressurized dispensing of certain products not previously possible. A pressurized container is inserted in the base of the tube or canister, pushing a plunger up into the substance held by the product. A valve is used to dispense the air from the canister.

SPACKLIT

DESIGNER/MANUFACTURER
Muli Bazak

DESIGNER BIOGRAPHY
Israeli designer Muli Bazak has a degree in Industrial Design from the Bezalel Academy of Art and Design.

CLIENT
Student project for the Bezalel Academy of Art and Design
www.bezalel.ac.il

MATERIAL
Thermoformed polymer and putty

MATERIAL PROPERTIES
Ergonomic, Malleable, Lightweight

Any tool employed by human hands qualifies as technology, even a tiny device for filling in a hole. The problem with most technological progress is that it fails to concurrently advance the state of design. But industrial designer Muli Bazak has managed both those feats with the kind of simple ingenuity that leaves one questioning how the old method lasted so long.

His project, executed while a student at the Bezalel Academy of Art and Design in Jerusalem, grew out of a quest for innovative solutions to familiar situations. "I started this project from the perspective of a student who moves frequently, which leads to the renovation of a lot of walls after removing a picture or a shelf."

The challenge with most spackling jobs is managing the goop itself. Spacklit resolves that with an ergonomic design that is flat on one side to sit flush with the wall and rounded on the other to be easily held, and squeezed, between two fingers. By merging packaging and tool into one, just the right amount of material can be delivered into the hole. Once dry, the bottom of the package can be used to sand down the surface. While to some it may seem little more than a handy pocket tool, its strength lies in its logic, a thoughtful sensibility holding true to the phrase "form follows function."

The base of the package is sandpaper, allowing the user to prepare the wall area. Then, by removing the tab from the back of the capsule and squeezing the body, the hole is cleanly filled with spackle.

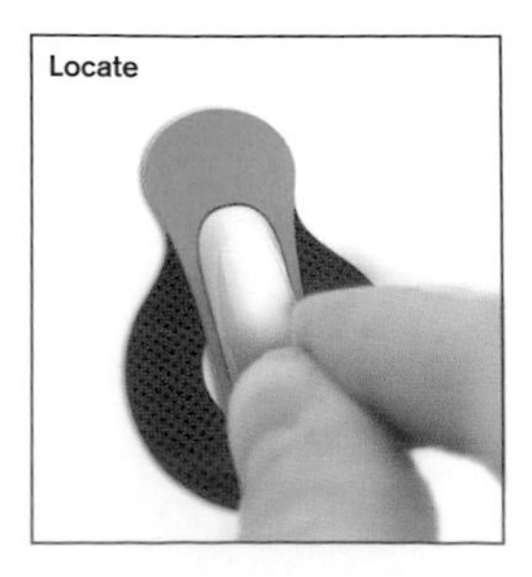

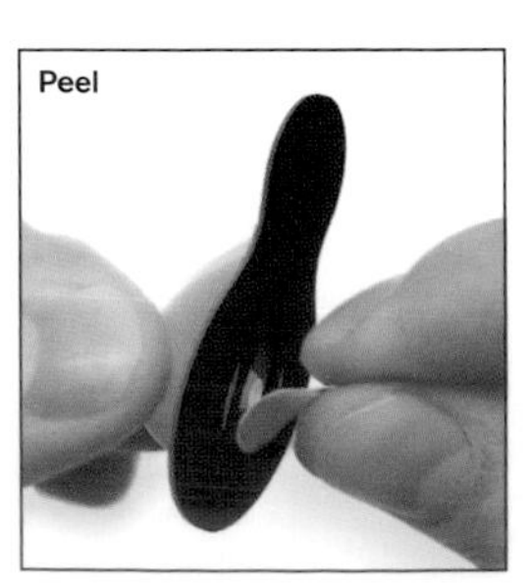

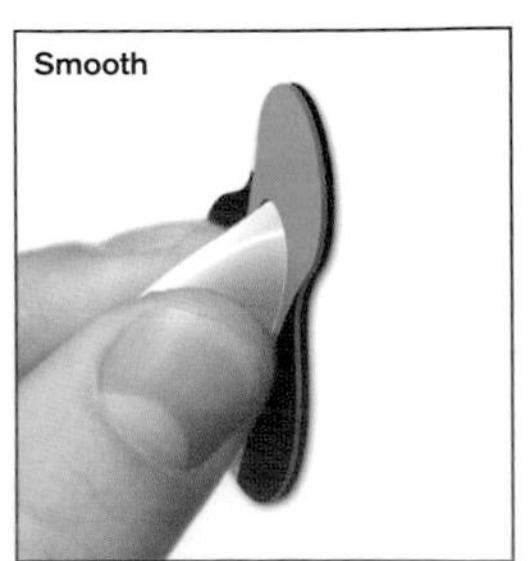

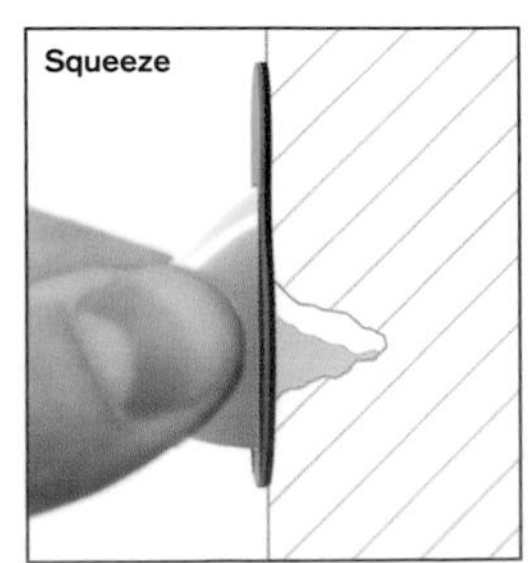

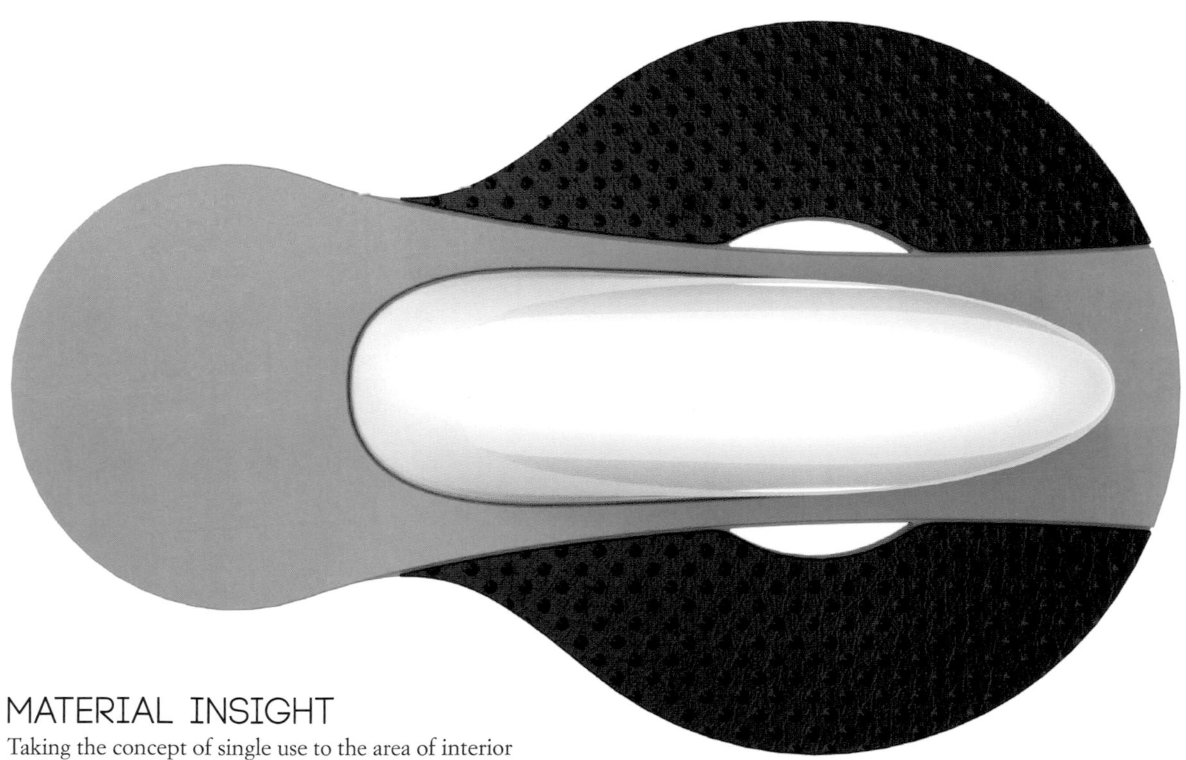

MATERIAL INSIGHT

Taking the concept of single use to the area of interior construction and repair, Spacklit potentially removes a lot of the messiness associated with minor repairs of drywall or plasterboard. The combination of an application-ready dose of moist putty or spackling compound in a moisture-impermeable sac with a small spreader enables holes to be filled and smoothed over in one step. A peel-away lid is removed when the product is to be used. The sac itself is a thermoformed clear polymer that acts as a small "handle" to be held between thumb and forefinger, which can be squeezed to deposit the correct amount of putty. The base plate used to smooth the surface is a semi-rigid molded polymer with sandpaper on its face.

Spackling wall imperfections typically requires multiple tools. Spacklit simplifies the process by melding ergonomic elegance with the efficiency of an all-in-one applicator.

BOTTOMS UP DRAFT BEER DISPENSING SYSTEM

DESIGNER
Josh Springer

DESIGNER BIOGRAPHY
Josh Springer is the founder and CEO of GrinOn Industries, the designer and fabricator of the Bottoms Up Draft Beer Dispensing System, based in Indianapolis, Indiana. The company distributes throughout North America and many parts of Europe.

MANUFACTURER
GrinOn Industries
www.grinonindustries.com

MATERIAL
Polymer cup and magnetic seal

MATERIAL PROPERTIES
Hands-Free, Efficient, Magnetic

It is hard not to smile at a company named GrinOn, especially when its product is a draft beer dispenser that fills beverages through the bottom of the cup. The idea for the business came to the owner in a daydream and was launched in that most auspicious of seedbeds—a garage. While the story may possess a folksy charm, one need only consider how far beer culture has progressed in the United States to appreciate that it is a thriving industry with a proud heritage that is not immune to innovation. In fact, beer culture worldwide is flourishing, from the proliferation of American craft brewers to countries such as Britain, Germany, and Belgium, which in recent years have been burnishing their own traditions.

All this makes the arrival of the Bottoms Up Draft Beer Dispensing System well timed. While thousands of restaurants serve great beers, many are staffed by people who know little about them, nor how to achieve a proper pour. The Bottoms Up cup and dispenser is a hands-free system that works to provide a pour calibrated to the brewer's recommended settings, maintaining quality and achieving the perfect one-inch collar of foam. Plus, it fills at a rate nine times faster than a traditional tap, less of an issue at the local pub but handy at a sports stadium, to which those waiting in line might readily reply, "Bottoms up!"

For those who love ale, draught beer delivers better texture and taste. But hand pours can be tricky and slow. The Bottoms Up dispenser achieves a faster and more accurate pour.

MATERIAL INSIGHT

As well as being simple, the concept also addresses a number of challenges to beer

dispensing in one go. The one-way valve that is the core of the innovation is based on a simple magnet at the base of the cup; this magnet is pushed up by a nozzle, enabling beer (or any other liquid) to be dispensed through the bottom, and in the case of beer controlling the amount of foam delivered, a major cause of wastage during normal pouring. The dispenser allows only a set amount of liquid to be delivered, meaning that the server can leave the cup there without fear of overspill. Once the cup is removed from the dispenser, the magnet re-engages with the base of the cup, creating a seal. The cup shape and the material it is made from need only be able to withstand the force of being placed onto the dispensing nozzle, and to contain the magnet valve.

As the cup is pushed down onto the dispensing tap, a magnetic ring-seal around a hole in the bottom lifts up and the cup fills with beer, hands free. Once it reaches capacity, it switches off. When the cup is removed, the magnet is pulled back into place, sealing the bottom.

ONE EIGHTY

DESIGNERS
Arttu Kuisma, Nikolo Kerimov, and Juho Kruskopf
www.motiff.fi

DESIGNER BIOGRAPHY
Based in Finland, Motiff is a creative studio formed by a group of friends with a passion for design, graduates of the Lahti University of Applied Sciences. Their houseware products, from shelving systems to tabletop items, offer a fresh perspective on familiar objects.

CLIENT
Concept

MANUFACTURER
Prototype (model protection pending)

MATERIAL
Paper-based fiberboard with a polymer coating

MATERIAL PROPERTIES
Malleable, Waterproof, Haptic

One Eighty's flexible yet rigid triangular pattern transforms the tube's shape every time it is used, creating a series of unique sculptural forms.

Materials can be clarifying. Rather than start with a design problem, begin with the material and then experiment. That is the approach, at least, that the young Finnish design firm Motiff took with One Eighty, arriving at a packaging solution that exploits the properties of fiber-based board coated with a film of plastic to act as a moisture barrier. The firm's material musings led them to find a subtle way of transforming the plain sheet into something distinctly sculptural.

By experimenting with crease patterns, they ultimately latched onto the functional and aesthetic possibilities of equilateral triangles. The result is a distinctive primary packaging design that has a triangulated surface, which operates like a diagrid (diagonal grid), producing a faceted, cubist-like silhouette.

Triangles by their very nature produce a dynamic effect, amplified here by the way the shape of the packaging changes every time it is used. When graphics are introduced, the faceted surface reads like a mosaic, a motif that can easily serve as the foundation for an extensive brand identity system. The full line of chromatically saturated tubes inspires with the possibilities inherent in basic geometry. Although meant to fit in the hand, the design has a panache that helps it stand out on the shelf and, as the designers suggest on their website, would attract someone who is likewise looking to stand out.

MATERIAL INSIGHT

The use of equilateral triangles on the surface of a flat plane with foldable creases at each edge creates the ability to distort the material without the typical deforming creases that can make a product look "used." The surface buckles along the edges of the triangles rather than randomly, giving an appearance of sculptural intention. The paper-based fiberboard used in this men's hair-cream tube has been

pressed to create fold lines within the surface. It has also been coated with a protective polymer film on both sides as a moisture, contamination, and soiling barrier, and printed with a graphic that mimics the lines of the triangles. The shape of the packaging changes every time you use it, so it provides an interactive user experience and stays interesting throughout its life cycle.

The creased pattern controls the behavior of the carton, making it possible to alter the shape without producing unpleasant wrinkles.

NO PACKAGE
SHIPPING VOLUMES
1. Shiping new phones
REDUCE: Minimize packaging, perhaps no primary packaging at all
REUSE: Same TRANSPORTPACK used several times (change interior placer?)
RECYCLE: Recyclable and compostable materials
RETHINK: no package?
2. In the shop
Buy only what you really need
SHOP
4. Recycle
Package and phone materials
3. Recycle shiping old phones
TAKEBACK: Same TRANSPORTPACK as shipment to POS
CO2
Reduce/lower transport costs and emissions
- Minimize weight
- less packaging
- no manuals
- not 1 charger per unit
- not 1 headphone set per unit
DESTRIBUTION CENTER AB
SE
SHOP
COMPONENTS
HOME
STAFF
CONSUMER

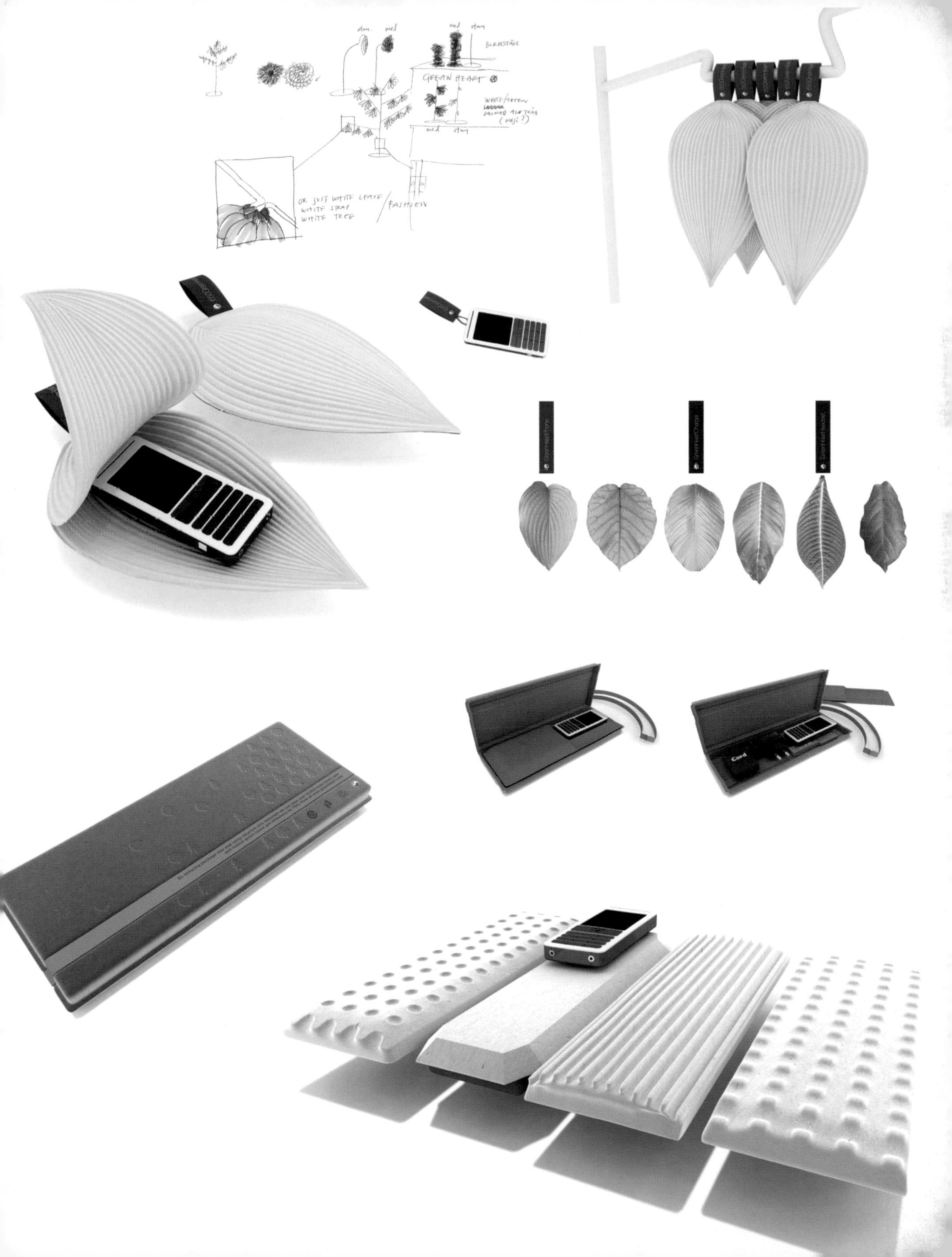
GREEN HEART
WHITE/GREEN
OR JUST WHITE LEAVE
WHITE STRAP
WHITE TREE
FASHION
Card

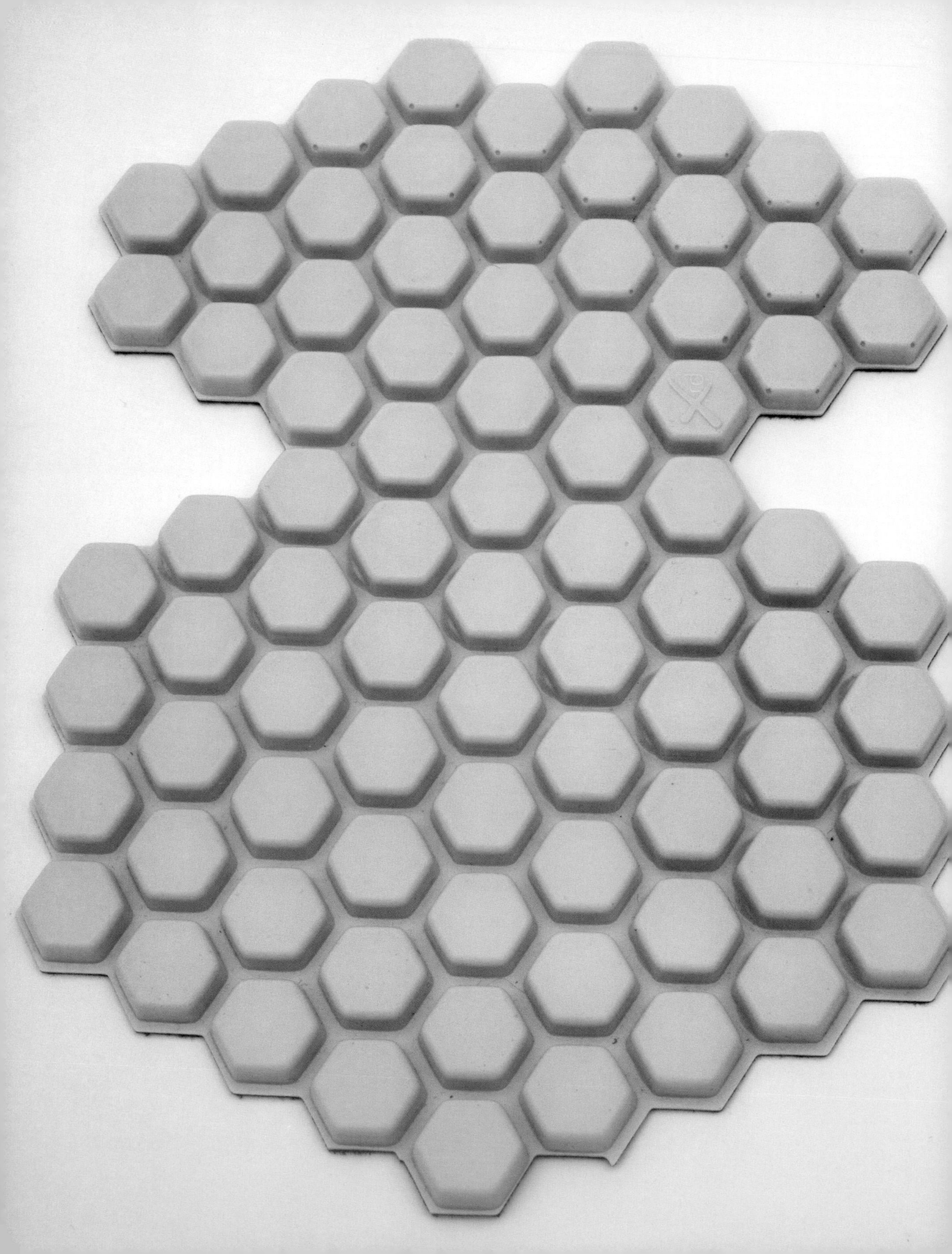

CHAPTER 4

ADVANCED PROTECTION

The protection in this case is really about safety, whether of the objects themselves or of those who come into contact with them; safety beyond the normal considerations of packaging design. As in Chapter 2, functionality trumps aesthetics; however, with this group of projects, the material and structural innovations have been created to solve one or more very specific challenges from the type of product they are to envelop. The concept of "advanced protection" in this chapter covers two particular areas of safety: one where the contents are of significant value or in some way delicate or fragile, and the other where the product is a danger to those who handle the package, such that in order to transport it safely, additional material or structural precautions are required.

For high-value items, at a basic level, this can mean layers of bubble wrap around the spout and handle of the expensive teapot you do not want chipped when delivered as a present. However, in order for it not to be entirely custom every time, and to build in a level of guarantee to the system, we must consider options beyond small bubbles of air sealed in polyethylene.

Structurally, highly elasticized rubbery films are now able to suspend the teapot free from impingement, and molded paper foams are able not only to protect from impact but also to provide good thermal insulation for products that need to be kept hot or cold. Structural innovations are also able to change the way we perceive certain materials. Pallets for industrial shipping can now be made from paper, provided that the sheets are folded and corrugated in the correct sequence.

opposite Poron XRD from Rogers Corporation offers a synergistic combination of advanced shock absorption and conformal flexibility, achieved through a PU foam encapsulated in a tough elastomeric skin, ideal for protective apparel.

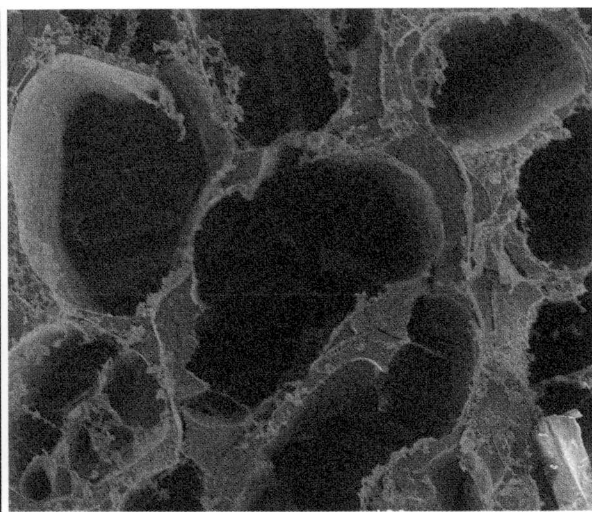

above left Compadre's AeroClay turns a mixture of clay and polymer into an open-cell, flame-resistant, lightweight foam whose impact and heat resistance make it a prime candidate for high-performance packaging.

above right Cells from the skin of a fish packaged using Mitsubishi Chemical's B-FreK conductive polymer films, which accelerate freezing so that smaller cells in the fish are undamaged in the process, improving the taste.

There is the use of materials "grown" specifically for the purpose, such as in the Bacs. packaging concept, in which protection is afforded in the way nature intended, through the use of foam-like cellulose that is built up around the product—in this case an egg—caused by bacterial work on glucose.

For hazardous items, there is sensibly a hierarchy of protection based on the level of danger. At one end, to ensure that fish is delivered fresh and without odor during transport, the number, material type, and order of multiple layers of plastic and metallic films are controlled to minimize weight and cost. Here the exact chemical reactions that occur in a recently dead animal must be considered—what gases are likely to be emitted and how are the color and structure of the skin and flesh to be maintained during chilling? Both ingress and egress of molecules through these films must be carefully controlled.

At the other end of this range is the packaging of materials that have been made toxic by their use and will remain dangerous for some time to come. Radioactive waste is of sufficient hazard that it needs not only to be entirely encased in glass to render it safe, but the glass containers themselves also need to be stored in geographical areas of minimal seismic activity in bunkers deep under the ground or within mountain caves.

Advanced protection can also mean security. And increasingly, the need to both ensure that the products and packaging are exactly what they claim to be (anti-counterfeiting) and also remain on the shelf until legally purchased is an industry in itself. Although the items protected tend to be of higher value, the packaging still needs to remain cost-effective, and the introduction of nanotechnology in the form of highly conductive inks and films is enabling this.

Through all these projects, safety remains the underlying theme: protection of the products from outside damage, whether physical or chemical, and also protection from the items themselves, since in our globally interconnected world there are now few things we use that did not travel from somewhere else to get here, and increasingly they require advanced protection to do so safely.

below Self-heating and self-cooling packaging by ScaldoPack relies on chemical reactions inside a pouch situated at the center of the liquid, which are activated by either shaking (for cold) or squeezing (for hot).

PEEPOO BAG

DESIGNER
Professor Anders Wilhelmson
Graphic design: Lowe Brindfors
www.wilhelmson.se
www.lowebrindfors.se

DESIGNER BIOGRAPHIES
An architect by training and a member of Sweden's Royal Academy of Fine Arts, Anders Wilhelmson invented the Peepoo solution in 2005 to help change the way people in poor and crowded communities contend with inadequate sanitation, one of the developing world's most intractable problems. Lowe Brindfors, part of the Lowe & Partners global network, is a Stockholm-based full-service communications agency.

CLIENT
Peepoople AB
www.peepoople.com

MANUFACTURER
Peepoople AB, Stockholm, Sweden (bag)
BASF (material)

MATERIAL
Low-density polyethylene (LDPE) bag with pathogen-killing urea

MATERIAL PROPERTIES
Low-Cost, Lightweight, Ergonomic

Peepoo was invented as a lightweight, low-cost, portable toilet that provides basic sanitation and impedes the transmission of disease in areas where these needs are ineffectively met.

Its whimsical name aside, the Peepoo bag means serious business. In developing countries, or in the aftermath of a natural disaster, where crowding, a lack of clean water, and the need for basic sanitation combine to increase the risk of communicable diseases, offering a way to defecate cleanly and privately is paramount. This is especially true for women and children, who are vulnerable to abuse when they must publicly attend to their personal needs.

To alleviate these conditions, the inventors of the Peepoo bag have sought to provide maximum hygiene with a minimum of material. Their sanitized, single-use toilet is made of a double-layer bio-plastic bag that sanitizes excreta, rendering contaminating pathogens inactive, not only as a barrier to the spread of such diseases as cholera and dysentery, but also enabling its use as a valuable fertilizer. Among the major benefits of this discreet, low-cost solution is that precious water resources are not needed, thereby helping to sever the contamination link between water and sanitation, a prime obstacle to healthy living.

The Peepoo further helps de-stigmatize a basic human need with a sprightly dose of ergonomic design. An inner bag unfolds to form a wide funnel that can be securely packaged once used by sliding the outer layer up and over the inner sac then tying with a knot. Rounded, sans-serif type sends an approachable, easy-to-read message. A green-and white-palette helps convey cleanliness and healthy growth, positive associations for the user, the environment, and Peepoo's larger mission—access for all to dignified and hygienic sanitation.

Pee
poo
Self-Sanitising Single-Use
Bio-Degradeable toilet
www.peepoople.com
Pee®
poo

A dual-layer design helps keep skin away from bacteria. The inner lining protects hands so they are clean for holding and closing the bag after use. The outer layer slides over the inner before it is tied into a knot for disposal.

The Peepoo School Program offers comprehensive training in hygiene and hand-washing practices to children, teachers, and staff. An image of Dubo the bear, the program mascot, is woven into the material.

MATERIAL INSIGHT

The collection of human excreta as a valuable fertilizer for crops has been practiced for millennia. The challenge has always been safely handling this valuable resource without transmitting deadly pathogens. The Peepoo bag, a plastic bag used as a one-time toilet in regions where sanitation is unavailable, provides just that. The low-density polyethylene (LDPE) bag contains 6 grams (0.2 ounces) of pathogen-killing urea that is activated as soon as it comes into contact with feces or urine. The human waste is broken down into ammonia and carbonate. As the urea breaks down, the pH value of the material increases and sanitization begins. Disease-causing microorganisms are inactivated within two to four weeks, depending on the surrounding temperature. The raw materials for this innovation are low-cost and lightweight enough that the bags can be delivered in the millions wherever they are needed.

With Peepoo, an affordable, sustainable, and scalable approach to safe sanitation is possible. Beyond trapping contaminants, it transforms human waste into pathogen-free fertilizer, providing a full-cycle solution for urban slums and refugee camps.

FRESHREALM VESSEL

DESIGNER
FreshRealm with RKS Design
www.FreshRealm.co
www.rksdesign.com

DESIGNER BIOGRAPHIES
FreshRealm is the brainchild of co-founder and CEO Michael R. Lippold and a diverse team. Before founding FreshRealm, Lippold spent over six years at Calavo Growers Inc. as Director of Strategic Development. Founded in 1980, RKS is a Californian-based design and innovation consulting firm whose goal is to deliver human-focused solutions with a global impact. It is a culturally diverse team consisting of researchers, strategists, engineers, and designers led by founder/CEO Ravi Sawhney and creative director/principal Lance Hussey.

MATERIAL
Corrugated polypropylene

MATERIAL PROPERTIES
Temperature-Controlled, Modular, Reusable

FreshRealm's system for delivering fresh food door-to-door begins with the Vessel, a reusable shipping container designed to work with existing mail-order carriers and eliminate wasteful, single-use packaging options.

Neither too warm, too cold, nor too rattled: These are the essential ingredients for shipping fresh food through the mail. Add to that the transport requirements of such carriers as FedEx and UPS, and the overlap between what is possible and what is perishable narrows significantly. In countries, including the United States, where people consume too much processed food because fresh alternatives, despite the proliferation of green markets, are not always available, FreshRealm's vision of a national network that allows everyone to benefit from better-quality food is ambitious and laudable. The key is refrigeration. "You can't simply throw icepacks into a cardboard box," explains John Styn, FreshRealm's iconoclastic co-founder.

In collaboration with RKS Design, a protective Vessel that would solve these challenges was developed. Intended for low-cost shipping through existing carriers and sturdy enough for reuse, the cube-shaped design contains an insulation system that keeps contents cold (no dry ice or toxic gel) and a set of modular drawers for use by an individual, a family, or—optimally, for reasons of cost—a group. Once empty, the Vessel can be sanitized and reused.

Eliminating spoilage is only part of the agenda. To improve distribution, every component is bar-coded and tracked through a cloud-based technology platform so that the entire chain of food suppliers, makers, and packers can efficiently work together. FreshRealm's system is a daring contrast to the current grocery system, giving new meaning to the phrase "connect the dots."

MATERIAL INSIGHT

The biggest constraint was keeping the temperature range between 0.3°C (32.5°F) and 4.5°C (40.1°F) for twenty-five hours in every inch of the food cargo area. This was achieved using an interior-molded polymer casing and door that

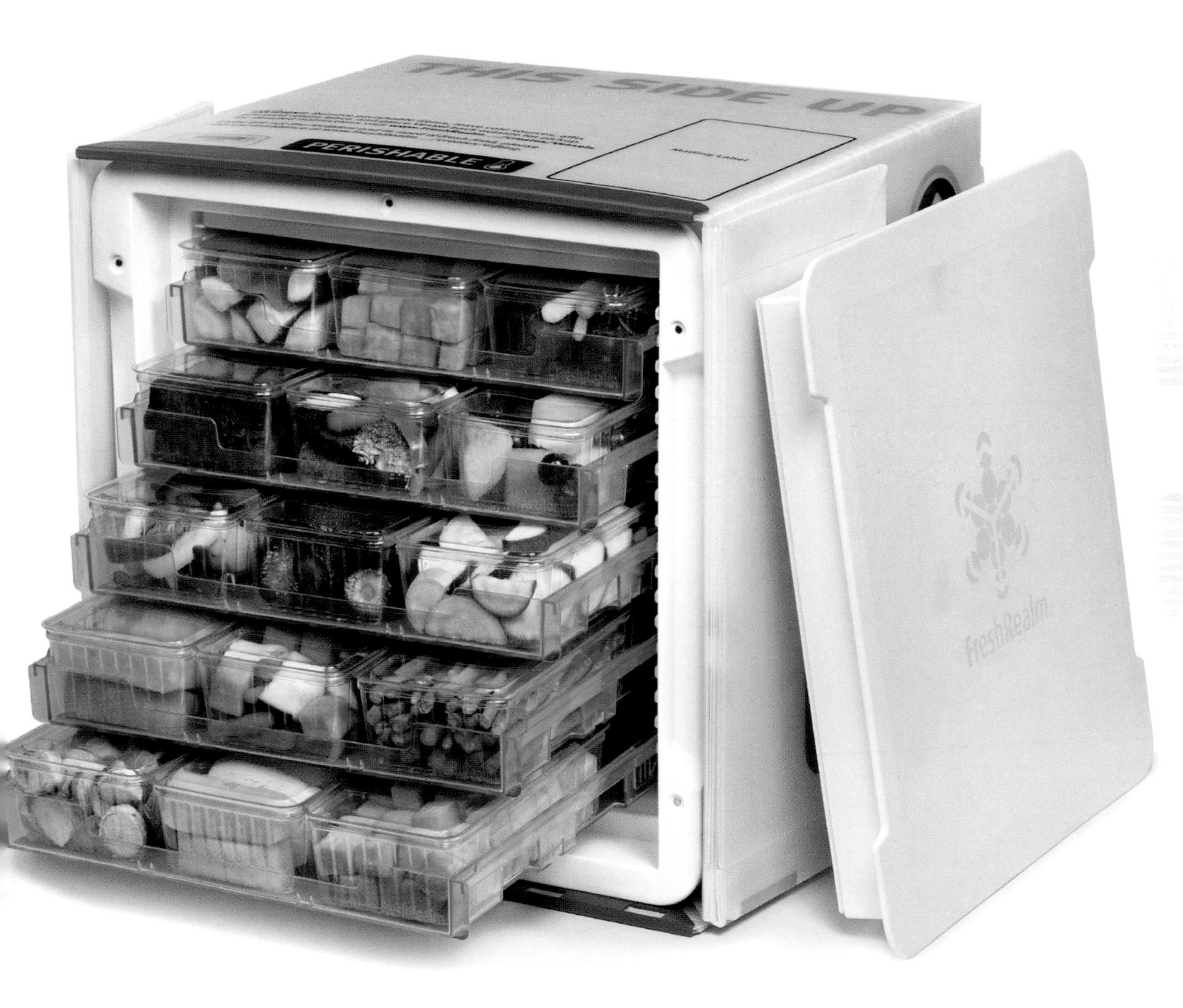

incorporates lightweight insulating polymer foam. The outer shipping box uses folded sheets of corrugated polypropylene that act as both insulators and impact-resistant walls to protect the case. The Vessel is delivered to a home full of food; the food is removed and the empty Vessel is picked up for sanitization and reuse. Every part of the packaging is barcoded: the corrugate, the cage, each drawer, and every food SKU (stock-keeping unit), enabling a fully traceable system that monitors food preparation, packing, and delivery status, and is integrated with the company's cloud e-commerce and inventory system.

Modular drawers maximize thermal properties to keep food chilled and allow for a single delivery to be easily divided up among several individuals or multiple families.

THE PAPER SKIN

DESIGNER
Geometry Global GmbH
www.geometryglobal.de

DESIGNER BIOGRAPHY
Part of the WPP network of communication firms, Geometry Global is the largest global activation agency of its kind, with 4,000 experts in fifty-six markets offering brand marketers proprietary strategies and creative ideas to influence consumer behavior and drive business growth.

CLIENTS
Fedrigoni and Leica
www.fedrigoni.com
www.leica-camera.com

MANUFACTURER
Carl Edelmann GmbH
www.edelmann.de

MATERIAL
Fedrigoni Constellation Jade packaging paper

MATERIAL PROPERTIES
Durable, Supple, Tactile

Instead of traditional saddle-colored leather, specially treated paper wraps the camera body in an elegant, pearlized surface. Its extraordinary properties allow it to withstand heavy abrasion, extreme temperatures, and corrosive liquids.

What might be arguably the most elegant camera ever made? It is not a trick question. But the answer is only partly about the device, and mostly about the material it comes wrapped in—paper, a substance that continues to astonish with its endlessly inventive possibilities and range of applications.

The camera is a Leica, the visual tool responsible for transforming photography in the early twentieth century from using unwieldy single plates to continuous film strips, and thereby the types of pictures that could be made. The paper is of equally impressive origins. Made in Italy by Fedrigoni, a manufacturer whose history dates back to 1888, the supple, pearlized "paper skin" shows that, in the hands of artisans, centuries-old materials can possess advanced properties and renewed life.

Leica cameras have traditionally been wrapped in leather, a saddle-colored banderole that is a hallmark of their design. For this limited-edition version, the exterior is covered in Fedrigoni's specially treated Constellation Jade paper, an ideal strategy for conveying the paper's exemplary toughness; the material is abrasion-resistant and can withstand extreme temperatures, from a bone-chilling -40°C (-40°F) to a wilting 70°C (158°F).

The secondary packaging is equally handsome. A box wrapped in pearl-colored paper functions like a book, with a special die-cut housing in the center for the camera. Silvery foil-stamped letters both subtle and bold showcase how Constellation Jade interacts with ink and other printing techniques. But it also reveals its supple response to three-dimensional objects, at the same time emphasizing its quilted, tactile appeal.

Leica
EDITION FEDRIGONI
LEICA
ELMARIT 1:2.8/24 ASPH.
4691363

To showcase Fedrigoni's beautiful yet tough product, a sample book-cum-box containing a limited-edition Leica X2 camera conveys the paper's assets, leveraging the lens-maker's heritage as an ambassador for the material.

MATERIAL INSIGHT

Effectively mimicking the durability and quality of the product it surrounds, the "paper skin" is a cellulose-based sheet that offers performance far above that of traditional papers, closer to that of leather in terms of abrasion- and chemical-resistance, as well as the ability to perform at temperatures that cycle from -40 to 70°C (-40 to 158°F). The material withstands all of Leica's extensive tests for the exterior body of the camera, yet does not lose its essential paper quality, also being used as the product's packaging. The multiple-colored sheets and exterior box structure are fashioned from the same material, and are sufficiently pliable to be folded, creased, and worked like regular paper as well as offering a highly receptive printable surface. The versatility of the material enables the packaging of the camera as well as all accessories to be made exclusively from Constellation Jade with no plastic at all, ensuring easy recyclability at end of life.

Leica
LEICA X2 EDITION FEDRIGONI

MUSHROOM® PACKAGING

DESIGNER/MANUFACTURER
Ecovative Design
www.ecovativedesign.com

DESIGNER BIOGRAPHY
While students at Rensselaer Polytechnic Institute, Eben Bayer and Gavin McIntyre discovered how to use fungi as rigid, molded materials. Encouraged to pursue their invention, they launched Ecovative, a material science company whose expertise is in home-compostable bioplastics based on mycelium. Today, Ecovative's team of scientists, engineers, product designers, and developers grows mushroom packaging for use by Fortune 500 companies. They have earned several awards, including the DuPont Packaging Innovation Award and Greener Package's Innovator of the Year.

CLIENT
Dell
www.dell.com

MATERIAL
Agricultural waste and fungal mycelium

MATERIAL PROPERTIES
Compostable/Biodegradable, Lightweight, Non-Toxic

Sensitive electronics require the utmost protection. Ecovative's design team collaborated with Dell's packaging engineers to "grow" biobased forms that protect fragile equipment.

The past decade has been a dynamic period in packaging for electronics. Personal computing devices at last disengaged from the notion that they needed to shout about advanced technology over a humanist understanding of communication, and benefited from a design epiphany, with no single tool—headphones, tablets, laptops, speakers—immune to the process.

Among those who introduced materially daring forms is Ecovative Design, a company committed to the development of a new type of compostable material based on mycelium, or mushrooms. Mushrooms, long considered in folklore and in botany to possess unique properties, are, as the team of engineers, scientists, researchers, product designers, and growers at Ecovative has shown, a high-performance and sustainable alternative to petroleum-based plastics and other harmful synthetics. The company's specialized indoor farming techniques allow them to grow mycelium in solid masses of different shapes and with custom-performance properties at prices that are competitive with conventional materials.

For Dell, one of the world's largest computer technology companies, shipping products configured to customer specifications is at the heart of their success story. But they are also intent on fulfilling a zero-carbon initiative, which relies on such materials as mushroom packaging to provide the utmost protection for sensitive electronic servers while in transit. Nothing, it would seem, is immune from nature as a wellspring for solutions intended to secure its own care and future health.

MATERIAL INSIGHT

These molded packaging profiles used to protect sensitive Dell server electronic components are in fact manufactured from fungi and agricultural waste, and are an alternative to expanded polystyrene (EPS). The pieces are created by

growing mycelium, a fungal organism, around rice husks, wheat chaff, and other discarded organic waste. The organisms grow into the full shape of the mold in seven days, forming miles of tiny white fibers that envelop and digest the seed husks, and bind them to form the final product. They need to be "baked" at a low temperature to inhibit further growth. The entire process uses about one tenth the energy per unit of material compared to the manufacturing of synthetic foams, and the material is completely back-yard compostable at the end of its life.

above and below Ecovative's mushroom packaging is grown from mycelium, a natural, self-assembling glue that digests agricultural waste to produce cost-competitive and environmentally responsible materials. This fully compostable solution challenges the widespread use of petrochemical foam insulation.

MODULAR VITRIFICATION SYSTEM (MVS)

DESIGNER/MANUFACTURER
Kurion Inc.
www.kurion.com

DESIGNER BIOGRAPHY
Co-founded by John Raymont and Josh Wolfe, Kurion creates technological solutions that minimize and stabilize nuclear and hazardous waste for safe and permanent disposal. Based in Irvine, California, the company offers separation, stabilization, and robotic technologies along with engineering and environmental services to help manage many of the world's largest nuclear and hazardous waste sites. In 2012, *The Wall Street Journal* honored Kurion with a Technology Innovation Award.

CLIENT
Various

MATERIAL
Ion-specific media that attach to radioactive particles and melt into glass

MATERIAL PROPERTIES
Rigid, Stable, Impenetrable

Packaging does not get much more critical than this. The "product" is deadly if handled or even brought close to humans and this hazard can remain for hundreds of years. Thus the challenge of how to package and dispose of a toxic substance such as nuclear waste is singularly daunting. The safe containment and disposal of radioactive material left over from producing either nuclear weapons (military) or nuclear fuel (civilian) means finding a fail-proof way to isolate it from the environment, typically by burying it deep in the earth.

To allow the radioactive waste to be moved and placed in an underground repository, it needs to be safely packaged for shipment and storage. To do this, it goes through a process of vitrification, a term rooted in the Latin word *vitrum*, meaning glass. The technology envelops contaminated material by heating it to melting temperatures then allowing it to cool, thereby forming a solid that traps dangerous particles and is easier to store.

Kurion, an innovator in waste management, is at the forefront of new solutions for treating and disposing of radioactive waste. The Modular Vitrification System comprises modular tanks that are used as the vessel in which the combination of glass and waste is mixed and heated, and are then allowed to cool and solidify to act as the shipping and storage containers. Each tank is considered sacrificial packaging and is removed from within the induction (heating) coils to allow a new container to be installed.

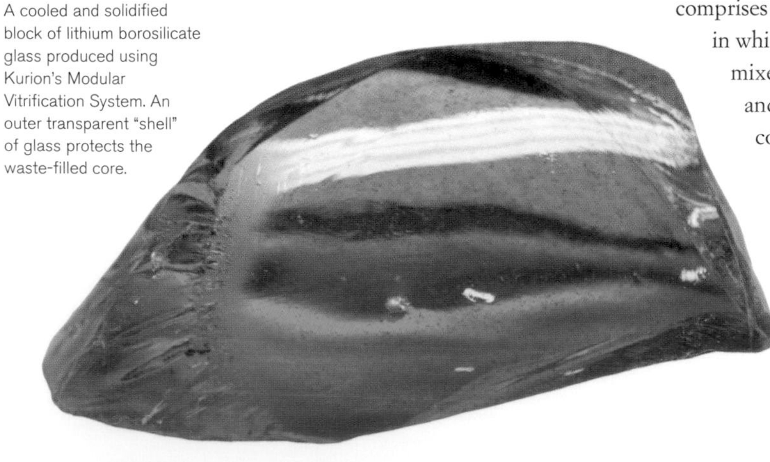

A cooled and solidified block of lithium borosilicate glass produced using Kurion's Modular Vitrification System. An outer transparent "shell" of glass protects the waste-filled core.

MATERIAL INSIGHT

Given that releasing it into deep space is prohibitive in terms of cost, energy, and safety, encasing radioactive waste in glass is still the most effective way of

Modular Vitrification System (MVS®)

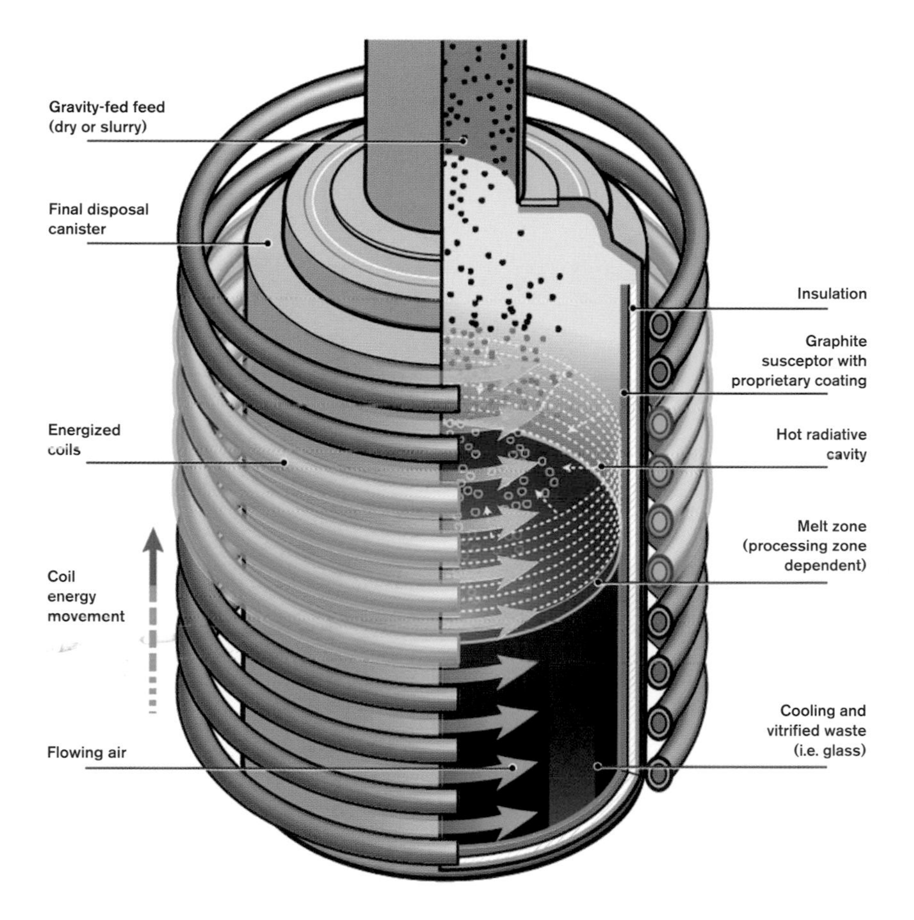

rendering it safe to be shipped and stored. Vitrification is the process whereby a material is changed into glass. This transformation, which can be done using waste glass cullet (glass that is crushed and ready to be remelted) or mineral-rich soil, is used to sheathe the waste material in molten glass of sufficient thickness to be impenetrable by even the most highly reactive materials. Typically used for liquid radioactive waste, the Modular Vitrification System simplifies and speeds up the process by using induction heating of specific containers that are filled with liquid waste. The containers are lined inside with graphite susceptors, which absorb the induction energy, heat the contents from the bottom up, and produce a vitrified block within the container that can be safely stored indefinitely.

During vitrification, Kurion uses inductive heating, where liquid nuclear waste is put in a chamber surrounded by coils. The result is a fluctuating magnetic field that produces heat without requiring direct contact.

CORRUGATED PAPER PALLET

DESIGNER/MANUFACTURER
Unipal International
www.unipalinternational.com

DESIGNER BIOGRAPHIES
For more than two decades, Bill Watson, Unipal CFO and acting plant manager, has worked in the pulp and paper industry, including finance, manufacturing, and supply-chain operations. Nick Wright, design and product development engineer, is the lead maintenance/operating technician for patented folding equipment, combining construction industry experience with commercial structural design and truss engineering. Terry Grier, production manager, provides hands-on shop-floor management.

CLIENT
Various

MATERIAL
Corrugated cardboard

MATERIAL PROPERTIES
Recyclable, Lightweight, Tensile Strength

Paper, these days, is a source of inspiration and a material of choice for everything from fashion to furniture, art to architecture. But it is also flexing its practical muscles by encroaching on the $10 billion-plus wooden pallet industry (think forestry), part of a global transport system that is responsible for moving most household goods from warehouses to stores.

As seaports look for cleaner, greener solutions and trucking companies look to make freight more compact, the pallet industry is evolving. Corrugated cardboard that has been folded, folded, and folded again until it possesses the strength of wood, yet is lighter and more flexible, has been increasingly viewed as a viable alternative.

Unipal International is at the forefront of corrugated-paper pallet manufacturing, and what it is out to do is no small feat. Its patented designs seek to resolve much of the skepticism surrounding cardboard versions—that they are flimsy, susceptible to moisture, and have a limited lifespan. To this end, Unipal's option is treated with a unique, water-resistant chemical and reinforced with stringers, beams that provide support in all directions. Even with those additions, their designs involve less weight, which means less fuel and emissions. Based on a modular system, a range of sizes can be constructed on-site without a hammer and nails, allowing for a greater amount—and a greater variety—of goods per shipment. Recycled and 100 percent recyclable, this is just one more indication of paper's tentacular reach.

The pallet weighs one third of a wood pallet, repels water, can be made from recycled cardboard, and is 100 percent recyclable.

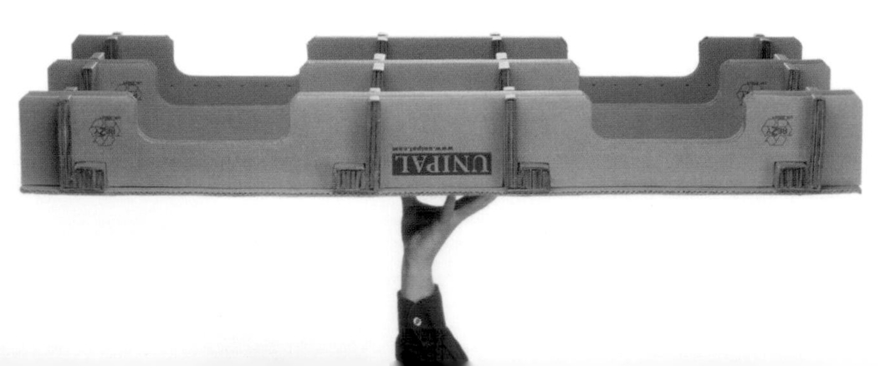

MATERIAL INSIGHT

These pallets, used for the large-volume shipping of products, are produced completely from standard corrugated cardboard. Using a patented folding process, the cardboard is scored, folded into a C shape and back on itself to create a five-layer laminate, each layer bonded to its neighbor. The resulting multilayer has exceptionally high bending strength and is typically coated with a water-resistant chemical, enabling it to be used in wet conditions. The folded panels are then die-cut to shape and combined with other shapes to create the pallet. There are various different permutations possible using the panels, depending on the size and format required. These pallets can carry dynamic loads in excess of 1,100 kilos (2,500 pounds). They can be recycled in standard municipal cardboard-recycling facilities and, unlike wood pallets, no nails need to be removed at end of life.

above and below Unipal corrugated technology offers several advantages over paper, wood, and plastic alternatives. Strong and lightweight, the frame requires no glue and thus can be easily disassembled for reuse.

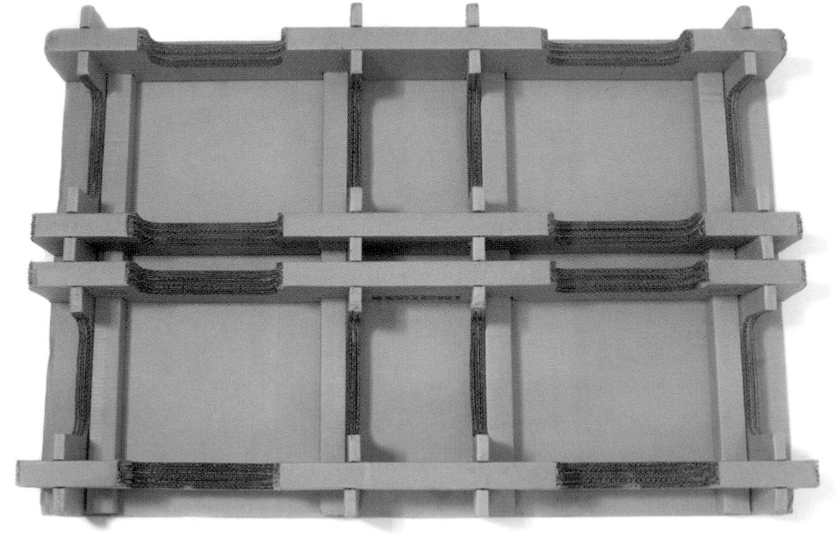

NATRALOCK WITH SIREN TECHNOLOGY

DESIGNER/MANUFACTURER
MeadWestvaco in collaboration with Vorbeck Materials mwv.inno-360.com
www.vorbeck.com

DESIGNER BIOGRAPHY
The collaborative team that designed, optimized, and commercialized the product consists of Mike Londo, Joel Forbes, David Miller, Michael Wade, and Mark Waddell from MeadWestvaco Corporation, Dan Scheffer and Sanjay Monie at Vorbeck Materials, and Scott Dowdell at Topflight Corporation.

CLIENT
Various

MATERIAL
Fully integrated conductive circuits, highly conductive graphene-based ink, and attachable, reusable module that sounds an alarm via an integrated speaker

MATERIAL PROPERTIES
Conductive, Printable, Portable

Flexible, strong and transparent, graphene is an efficient conductor of heat and electricity, allowing it to push the performance of inks and coatings for printed electronics.

Theft? In addition to brand identity, structural design, graphics, shelf presence, sustainability, and shipping, packaging designers also need to outsmart thieves? At MeadWestvaco (MWV), a global packaging company, the answer is a decisive yes.

While retail design has become more sophisticated, so has thievery. The goal in an era of nanotechnology, however, is for the security to be imperceptible so as not to inhibit a customer's engagement with a product. Few shopping moments are more dispiriting or demotivating than attempting to try on a sumptuous piece of apparel, only to discover it is chained to a display fixture.

To overcome this, MWV worked with Vorbeck to develop Natralock Security Packaging with Siren Technology. As the name suggests, Natralock is made from sustainable paperboard (a major improvement over plastic clamshells), which contains an extra layer of security via a graphene-based conductive ink. If a package is tampered with or opened and the product removed, the anti-theft system is activated and the package sounds an alarm. The siren is made possible with graphene-based conductive ink from Vorbeck Materials.

"Siren Technology is the next step in security packaging," says Joel Forbes, innovation commercial team leader at MWV. "With conductive inks creating circuitry within the packaging, the product is protected against tampering and theft. Products no longer have to be locked up, allowing for open merchandising and increased sales."

MATERIAL INSIGHT

This is probably the first commercial application in packaging of graphene, the one-atom thick, highly conductive wonder material that could revolutionize materials science. It is used as the conductive part of the ink (called Vor-ink) that is printed in specific patterns

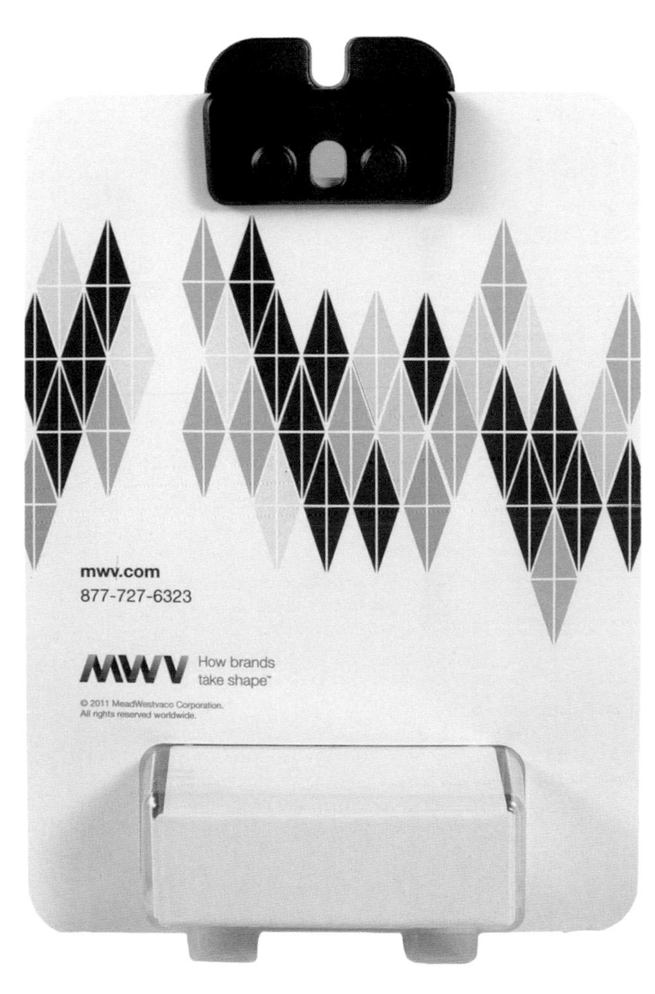

across the surface of the packaging product. It forms a grid that when disrupted—by tearing, cutting, pulling apart, or leaving the store—breaks the circuit connected to the power pack, emitting a loud siren to alert store staff. Graphene was used because it has a conductivity level not achievable by other carbon or polymer inks. The "siren" is produced by the large black box affixed to the top of the packaging, which also contains the power and connectors to the printed paper. There is an advantage to its large size, visually deterring potential thieves through clear indication that this package is not to be tampered with.

Today's electronic smart packaging includes anti-theft devices strong enough to trigger an alarm when a thief attempts to steal or tamper with a product yet small enough not to deter shoppers.

BACS.

DESIGNER/MANUFACTURER
Mareike Frensemeier
www.researchgate.net/profile/Mareike_Frensemeier/

DESIGNER BIOGRAPHY
After studying Biomimetics, Industrial Design and 3D Mobility, German-born Mareike Frensemeier earned an MSc in Advanced Material Science and Engineering from the University of Saarland/ETSEIB in Barcelona, Spain. She went on to pursue her PhD in Functional Surfaces, Metallic Microstructures and Smart Materials at INM (the Leibniz Institute for New Materials) in Saarbrücken, Germany.

CLIENT
Concept

MATERIAL
Bacterial nanocellulose built from glucose and water by *Acetobacter xylinum* (acetic acid bacteria)

MATERIAL PROPERTIES
Dynamic, Organic, Customizable

By feeding the bacteria with sugar, a biodegradable paper-like shell is formed around an object, encasing it in superfine fibers that offer high strength and customizable applications.

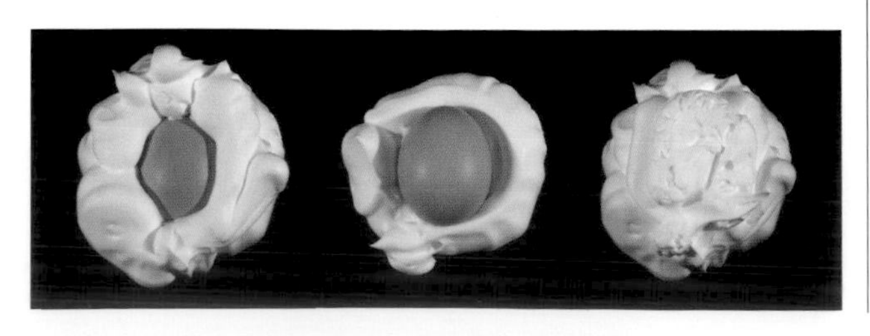

Innovation is not always about the big picture. An education that blends biomimetics and industrial design with engineering and materials science inspired the PhD candidate Mareike Frensemeier to become familiar with the most diminutive of characters: bacteria. By harnessing the unique properties of the acetic acid bacterium *Acetobacter xylinum*, she is able to convert glucose into super-fine, super-tough cellulose fibers for use as packaging. The fibers can grow to varying thicknesses and shapes, even—when coated on the surface of an object—encasing it in a biodegradable shell.

Like a child gazing at a hidden world, Frensemeier conveys a fascination with the microscopic living organisms that are her producers. "To analyze nature's output, a result of millions of years of the strictest optimization technique—survival of the fittest—is to encounter the largest known pool of long-range, fully proofed materials and mechanisms," she says, suggesting that her studies, while science-based, have also acted on her passion as a channel for deep thinking.

Frensemeier's research has led to textural variations of nanocellulose that can be adapted to different product needs: A hydrogel is suitable for fresh food, paper fulfills hygienic needs, and a foam version delivers advanced protection. An innate UV protection further extends its value to light-sensitive products, such as film and photo papers or the active ingredients in health and beauty products.

MATERIAL INSIGHT

This excellent example of a creative use of bacterial action provides a unique solution to concerns about packaging waste. The bacterium is called *Acetobacter xylinum*. It will grow quickly at room temperature to create a cellulose hydrogel, paper, or foam structure. The foam is actually a network of tiny, strong fibers on a nanostructured scale,

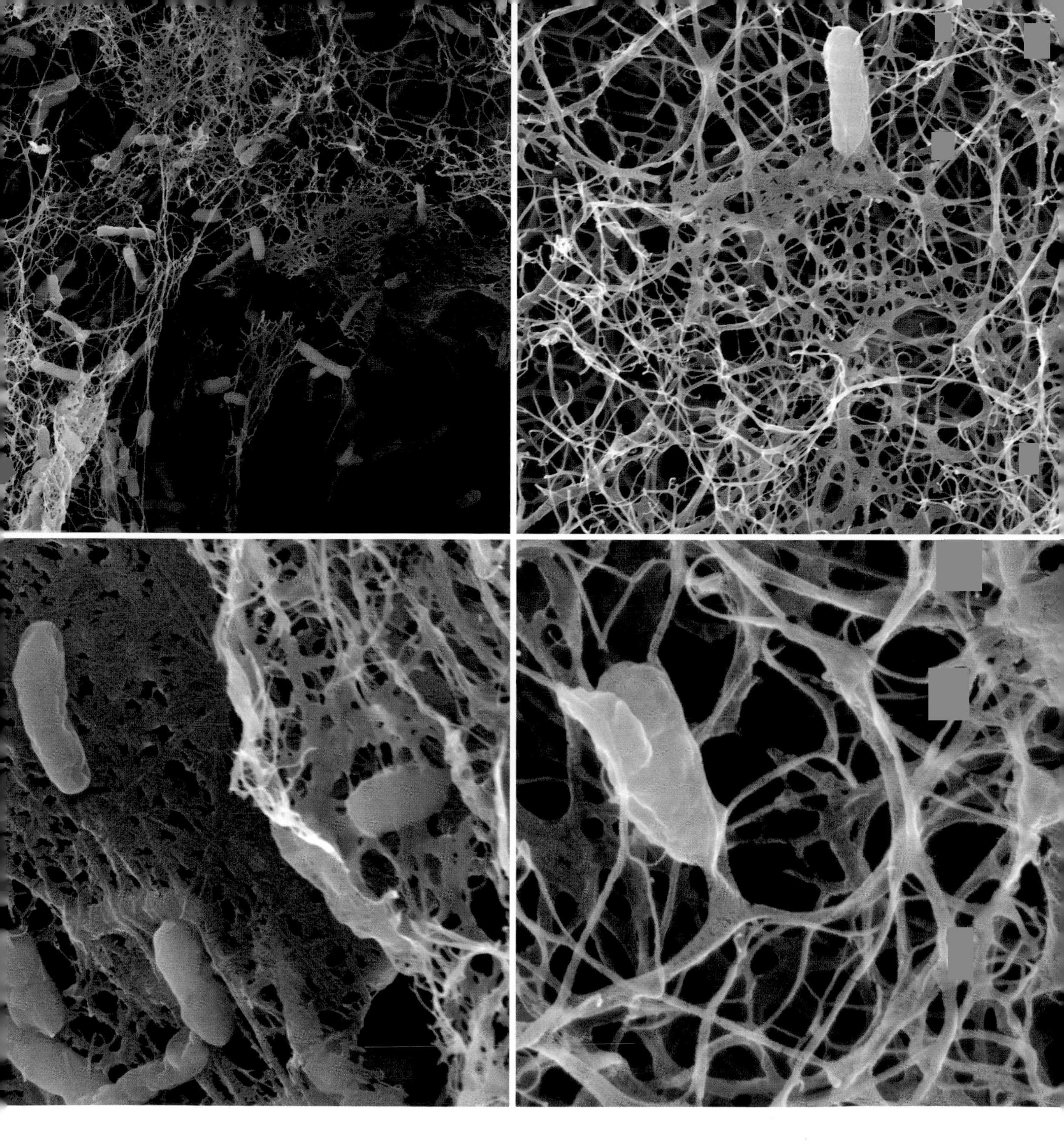

which grows from a glucose–water solution. It can be grown into any shape, such as a box or barrel, or directly grown on a surface, self-assembling around it to create a lightweight, paper-like foam that protects the object being packaged from impact, and to a certain extent from water and air ingress. The foam is electrically and thermally insulating, and is completely biodegradable at the end of its useful life.

Energy-efficient and eco-friendly, Bacs. exemplifies the exciting potential for complex packaging challenges to be met through the application of design thinking.

NATURALLY CLICQUOT

DESIGNER
Cédric Ragot
www.cedricragot.com

DESIGNER BIOGRAPHY
French industrial designer Cédric Ragot has collaborated on consumer electronics, tableware, furniture, and sports equipment with such clients as Bitossi Ceramiche, Artuce, Swarovski, Krups, Roche Bobois, and Häagen-Dazs, among others. His work has been exhibited around the globe and his many honors include the 2014 Fragrance Foundation Award for Best Design, a 2012 Red Dot Design Award, and a 2008 *Wallpaper** Design Award.

CLIENT
Veuve Clicquot
www.veuve-clicquot.com

MATERIAL
Potato starch and paper

MATERIAL PROPERTIES
Lightweight, Isothermic, Biodegradable

Initially it is a stretch to imagine the humble potato as the origin of the next phase of innovation for a luxury brand such as Veuve Cliquot champagne. A portable cooler designed by Cédric Ragot that conforms to the shape of the bottle and opens like a clamshell has been crafted from potato starch and recyclable paper. The result is a 100 percent biodegradable isothermal container that keeps the bottle cool for up to two hours. The white exterior suggests the cooling effect of ice and snow, while providing a pristine backdrop for the appearance of the deep-green bottle and distinctive orange label. A rounded handle attached to the neck mimics the arc made by the sparkling wine's celebratory spray, while suggesting the grab-and-go ease of a picnic.

"Act with audacity," the Widow Clicquot advised a great-grandchild, a prescient approach that helped her transform the winery she took over on her husband's death into a renowned commercial empire within her own lifetime. Evolving a luxury brand, especially the packaging, requires prudence and care if fresh design elements are to successfully connect with existing customers while capturing the attention of new prospects. Expansion involves the careful distillation and modernization of an established design language, a visual skill intended to fulfill the consumer's continued desire for something special and unique.

MATERIAL INSIGHT

Presenting a quirky take on what is essentially a thermal insulation container for bottled champagne, this molded-fiber-based packaging offers a lower-environmental-impact alternative to expanded polystyrene. The form is created

The portable packaging for this elite champagne is made from potato starch and recycled paper, which together possess isothermal properties.

from a combination of waste by-product starch—which acts as a binding resin—and cellulose fibers, which give the shape its strength. It is formed by pressing a wet mixture of the two materials in a heated tool. The composite mixture is biodegradable and potentially recyclable, but is durable enough to be reused. It is also lightweight and resists impact (it can be dropped without the glass shattering), and the design of the carrying handle means that less heat is transferred through the container, keeping the champagne chilled for more than two hours.

The form of this clamshell cooler is typical of the clean, sculptural approach of its designer, Cédric Ragot.

INTELLIGENT PACKAGING PLATFORM

DESIGNER/MANUFACTURER
Bemis Performance Packaging and Thin Film Electronics ASA
www.bemis.com
www.thinfilm.no

DESIGNER BIOGRAPHY
Thin Film Electronics ASA is an innovator in the field of printed electronics. It has been at the forefront of commercializing printed rewritable memory, and now creates printed system products with such capabilities as memory, sensing, display, and wireless communication. A global, publicly traded company, its headquarters are in Oslo, Norway.

MATERIAL
PDPS (printed dopant polysilicon) technology that enables high-performance transistors to support high-frequency RF circuitry.

MATERIAL PROPERTIES
Responsive, Customizable, Cost-Effective

Printed electronics are gaining traction as a way to create wafer-thin electronic and optical components that can collect and communicate information on a range of surfaces. "I magine the transformative effect across industries if printed electronics and wireless-enabled smart sensor labels could bring *just enough* intelligence to trillions of everyday objects," says Dr. Davor Sutija, CEO of Norwegian printed electronics developer Thin Film Electronics ASA, in a blog post for *Innovation Insights*. The advantages are especially clear for packaging, particularly when the contents are perishable. Tracking information wirelessly about freshness and environmental conditions on food, pharmaceuticals and personal-care products has wide-ranging economic and public-health benefits.

A collaboration with Bemis, a Fortune 500 supplier of flexible packaging and pressure-sensitive materials, gives both companies an edge. For example, circuit- and sensor-embedded labels—strips of clear film printed with electronically conductive ink that can withstand pressure, heat, and moisture—can be integrated to transmit customized data, paving the way for "packaging that talks." Since different product categories have different sensing needs, customization of smart packaging is key.

It is widely known that multiple factors influence purchasing decisions, including convenience, quality, and reliability. So even a small amount of sensing capability can have an impact. Ultimately, what smart packages convey will allow manufacturers to be more responsive and consumers to make more informed choices.

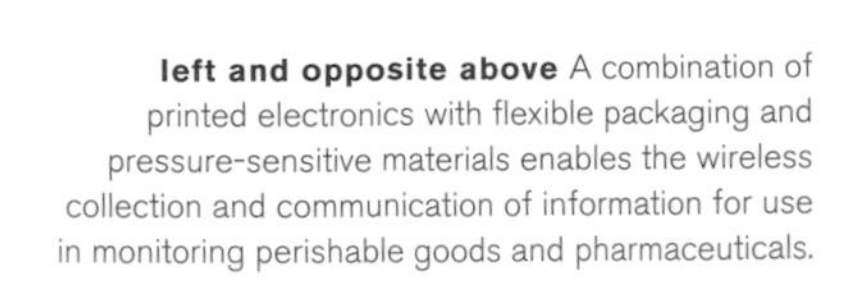
left and opposite above A combination of printed electronics with flexible packaging and pressure-sensitive materials enables the wireless collection and communication of information for use in monitoring perishable goods and pharmaceuticals.

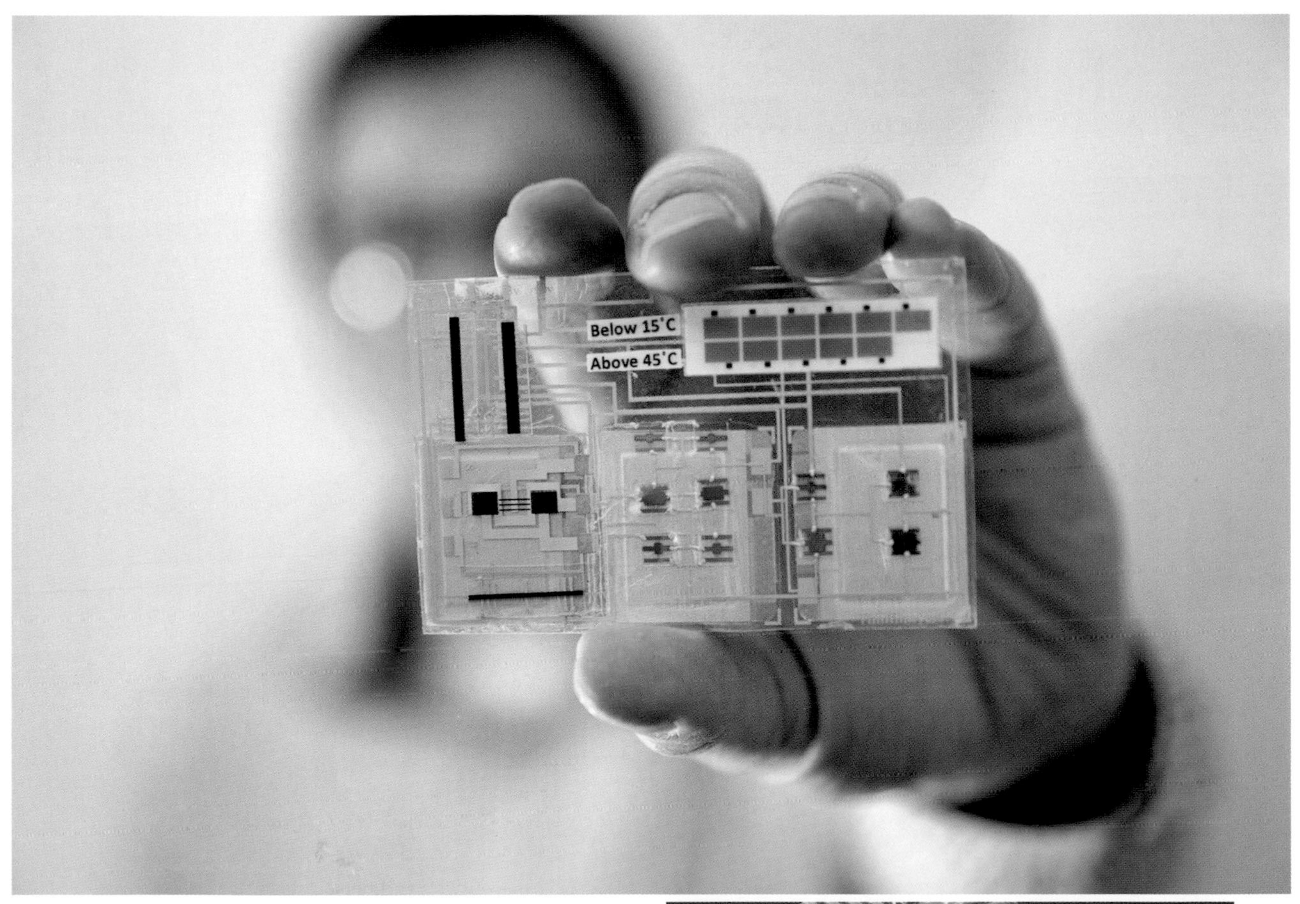

MATERIAL INSIGHT

This intelligent packaging tag is the industry's first integrated-system product to combine printed electronics technology, real-time sensing capability, and near-field communication (NFC) functionality. The Thinfilm smart label platform is designed to accommodate a variety of sensing elements, both printed and conventional, including temperature, humidity, mechanical shock, blood oxygen levels, and glucose levels. Depending on the application and necessary components, labels may either be fully printed or feature a combination of printed and surface-mounted elements. The company's PDPS (printed dopant polysilicon) technology is based on a hybrid manufacturing process that leverages print methods in key process steps. The technology enables high-performance transistors that support the high frequency RF circuitry required for NFC communications in these printed labels. Because the smart label contains a unique ID in addition to the sensor data, it is possible to log the alert in a cloud-based application for further analysis.

The Intelligent Packaging Platform can monitor and record key physical properties and environmental data for a wide range of products, from meat and cheese to medical devices and personal care items.

EXPANDOS

DESIGNER/MANUFACTURER
FoldedPak
www.expandos.com

DESIGNER BIOGRAPHY
Denver, Colorado-based FoldedPak developed ExpandOS in an effort to solve shipping issues for a variety of customers all over the world seeking an environmentally friendly packaging material made from sustainably manufactured paper that is also 100 percent recyclable. Led by CEO Jeff Boothman, the company drew on five decades of experience in packaging and transportation for its protective packaging solution.

CLIENT
Various

MATERIAL
Cardboard strips folded into triangles

MATERIAL PROPERTIES
Recyclable, Lightweight, Printable

An avalanche of paperboard pyramids tumbles into a box—light, three-dimensional, interlocking against the carton's rigid sides. Products destined for shipping are nestled into this protective cocoon, and the void filled in with the same shapes so that the product, no matter how delicate, is immune to the punishing effects of being dropped, crushed, or thrown.

Although not yet the cushioning material of choice against such default options as bubble wrap and Styrofoam peanuts, ExpandOS (short for Expand On Site) packing filler serves in the world of shipping as an emblem of modesty, intelligence, and inventive perseverance. Whether or not something so simple as a punched-out piece of paperboard can rid us of petrol-based packing materials at a time when we desperately need sustainable alternatives may be an unknown quantity at the moment, but its speed and efficiency in filling the space around any object lends itself to such a hopeful supposition. Indeed, ExpandOS delivers remarkable shock-absorbing capabilities yet is entirely recyclable.

Inventor William Oliver employed serrated edges to allow the pieces of cardboard to lock together, resulting in a sturdy matrix that has been called "paper cement." Circular holes have been punched into the center to reduce weight and allow further openings for pieces to engage. Among the material's bonus features is that it can be printed with a logo or graphic message.

A network of circular holes helps to minimize weight and enables the elements to further interlock.

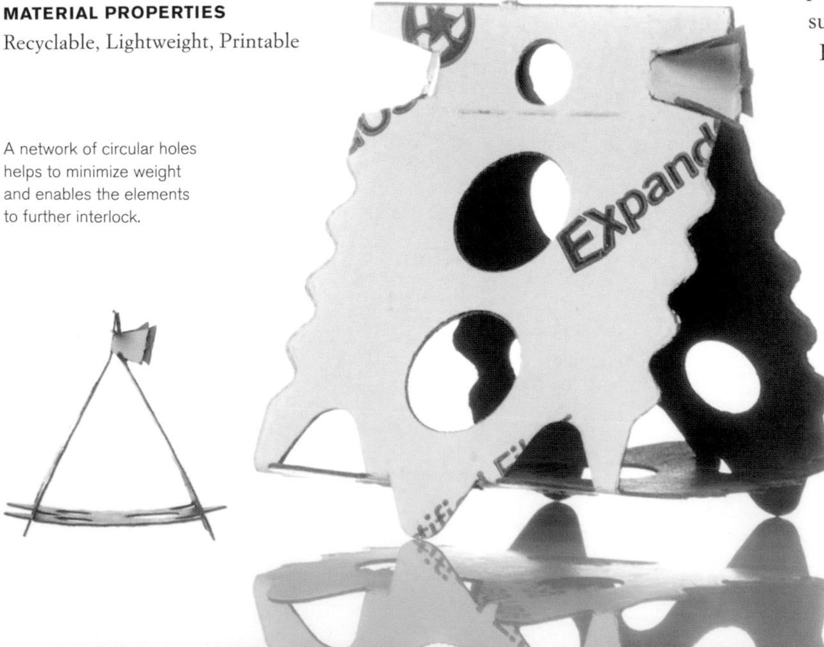

MATERIAL INSIGHT

This 3D interlocking packaging material, created from sheets of cardboard, is a recyclable, biodegradable, and non-toxic alternative to traditional packing materials such as foam-in-place and bubble wrap. Using an expander machine, sheets of fan-folded, die-cut fiberboard paper are used to create lightweight, pyramid-shaped elements without glue or adhesives. These pyramids have a series of perforations that reduce weight and allow a rapid flow into packages, seeking tangential surfaces so as to create an interlocked protective layer. The material is shock-absorbing, is completely recyclable after use, and takes up a twenty-fifth of the storage space needed for the expanded product. The packaging material is shipped flat in bundles with a size of 47.5 x 30.5 x 24 centimeters (18.75 x 12 x 9.5 inches) and can be printed with logos or other information. One bundle will fill twenty-five-cubic-foot (28.3-liter) boxes.

The pieces work together as a unified system to suspend the product, absorb shock, and reduce movement during shipping, eliminating the need for additional wraps, sheets, tape, or dividers.

FRAGILE

DESIGNER
Mireia Gordi i Vila, with support from the Victoria & Albert packing team
www.cargocollective.com/please_draw

DESIGNER BIOGRAPHY
Mireia Gordi i Vila is a product designer from Barcelona based in London. Before attending the Royal College of Art, she worked as a research and development designer in France, and as a product and exhibition designer in the Netherlands.

MANUFACTURER
Concept

CLIENT
Graduate show at the Royal College of Art, London
www.rca.ac.uk

MATERIAL
Elastic, composite membrane

MATERIAL PROPERTIES
Flexible, Reusable, Modular

Evidence shows that, with the right encouragement, new ideas spur more new ideas. This is valuable information for anyone looking to cultivate an environment where innovation is possible. To tap into fresh thinking, research shows that asking the right questions is more important than having the right answers.

Such was the starting point for Barcelona-born Mireia Gordi i Vila's Fragile packaging system. While her field of inquiry was shipping, her questions focused on materials: How can one get the best performance with a minimum amount of material? Explorations into the possibilities of prosthetic silicone and polyester power mesh, neither of which on its own was sufficiently strong, cushioning, or elastic, showed promise when the two were combined. Together they form a thin, pliable membrane that, when stretched around a stable frame, provides a supportive skin around valuables and breakables, creating a supple environment during transport akin to the net used by trapeze artists. Encased between two sheets of the membrane, the contents easily rebound from jostling and vibrations.

Optimally, the system can be reused for objects of different shapes, weights, and sizes. The dimensions of the frame allow for breakables that fit in hand luggage all the way up to shipping crates. The result is a greatly needed alternative to bubble wrap and Styrofoam peanuts, a lightweight and elastic passport to travel that began with a mere "what if...?"

An elastic composite membrane traps objects of different shapes and weights, producing a standard system for non-standard items.

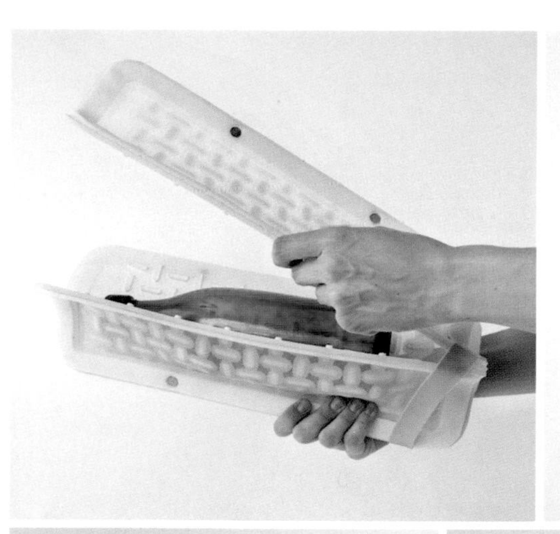

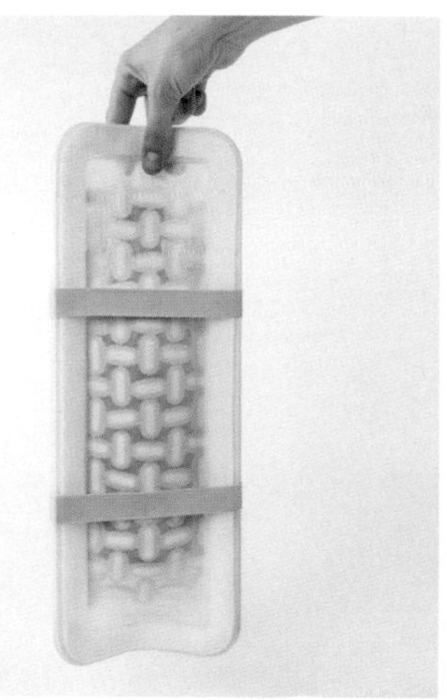

MATERIAL INSIGHT

This simple solution for packaging fragile items relies on the highly elastic nature of an elastomeric polymer film. Elastomers—synthetic alternatives to rubber—are a class of plastic that combine on a microscopic scale stretchy rubbers and more rigid plastics to give tough, tear-resistant films capable of stretching up to 1,000 percent repeatedly without loss in performance. Stretching this membrane within a frame and placing the object to be packaged between two of these frames suspends the object safely during transit. The membrane can be solid in format, or mesh for reduced weight and greater breathability or visual recognition. Vibration and impact are absorbed by the elasticity of the film without its ever touching the frame or the other outer surfaces of the box. The packaging can be reused multiple times, and the shape of the object is irrelevant so long as it fits within the confines of the frame.

top A specially designed frame for bottles suspends the object inside the reusable tensile membrane, then slips easily inside a standard poster tube (**bottom right**).

above Super-thin layers of prosthetic silicone and polyester power mesh provide the right symbiosis of grip, cushioning, and protection.

FROM:
TO:
TEENAGE ENGINEERING
TEENAGE ENGINEERING OP-1
MESSAGE:
TE
002.AS.809
7 350073 030416
BOX 1: 150*150*120MM BOX 2: 138*317MM
teenageengineering.com

10代工学
teenage

above Printoo, an Arduino-based platform of paper-thin, low-power boards and modules, gives designers new levels of creative flexibility. The primary core processor connects to other displays, power, switches, and hardware.

opposite above Printoo's Electronic Vote, an interactive digital voting system. The tally of votes on the left is created by an electrochromic display powered by a coin cell battery and the voting "buttons"are coated in conductive paint.

opposite below "Printoo Man," a simple display module, can be constructed with simple coding, using a Printoo Core module, an LED matrix (8×8) display, and a coin cell battery holder.

CHAPTER 5 INTERACTIVE

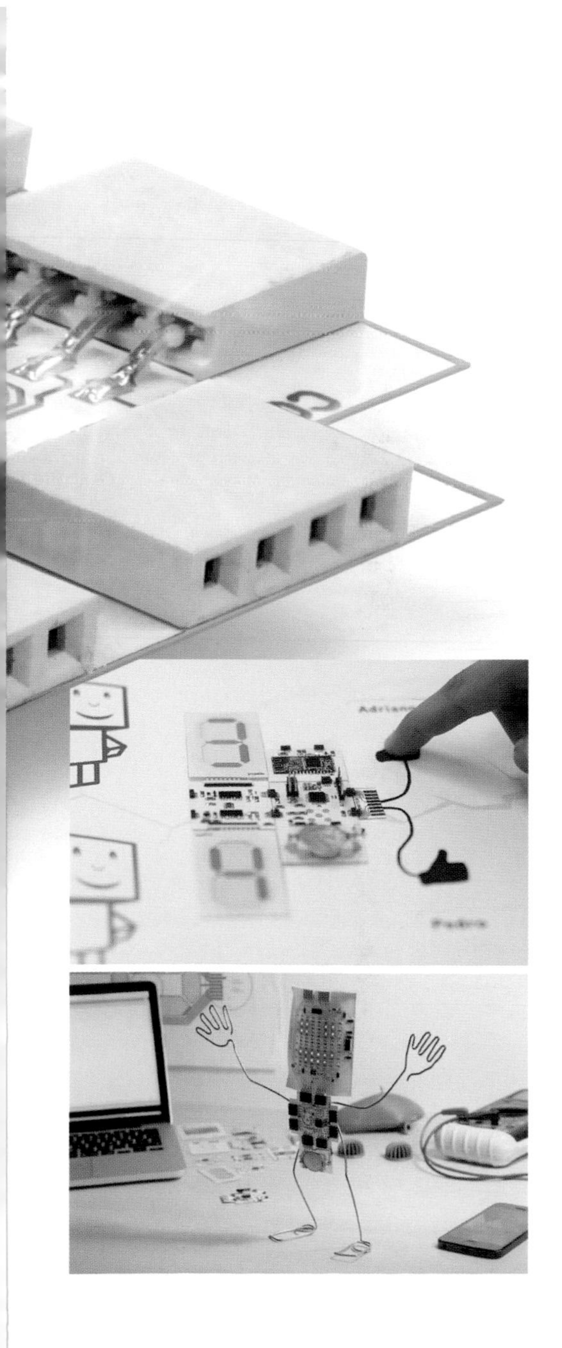

It is probable that in the not-too-distant future, almost all mass-produced packaging will include some form of interactive function. Whether fun and engaging for the consumer, such as a light-up panel that lets you know of a new discount, or essential to the efficient tracking of the product and thus invisible to you or me, intelligence will be built in.

This "interactivity" includes RFID, sensors, batteries, screens, touch-sensitive surfaces, and memory storage devices that can learn and update information about the product, as well as location trackers (GPS) and exposure meters (cold, air, moisture, specific gases such as CO_2, N, NO_2, and ozone). These types of devices are readily available, accurate, and small, but the challenge is making them viable for packaging. Since the majority of packaging is intended for short-term use, these solutions need to be lightweight, achieve high volumes, be exceptionally low-cost, and preferably have the potential to be recycled in some way at end of life.

The method by which the majority of these intelligent packages have become more mainstream and begun to be used in packaging is printing. The ability to print using high-speed methods such as gravure (flexible packaging), offset (newspapers), or flexo (retail shopping bags) brings down the cost for production at large volumes, required for many packaging sleeves, boxes, cartons, and films, achieving piece costs of fractions of a cent. The sensors themselves tend to be a combination of multiple layers of specific polymer, metal, and metal salt to create the circuits needed to sense or power, or simply to act as a wire-like connector

HUGO RED

DESIGNER
P&G Prestige design team, Sparkle Design Agency—Open D Group
www.prestigeproducts.hu
www.sparkle-design.com

DESIGNER BIOGRAPHY
David Wright is a principal researcher with the P&G Prestige packaging team, based in Surrey, UK. He has worked at Procter & Gamble since 2001, and has extensive experience in the Prestige fragrance category, with a core expertise in injection molding and complex decoration. He was part of the P&G team that designed the Hugo Red bottle.

CLIENT
Hugo Boss
www.hugoboss.com

MANUFACTURER
Glass bottle by Heinz Glas, glass decoration by Dekorglass
www.heinz-glas.com/en
www.dekorglass.pl

MATERIAL
Bottle coated in red thermochromic ink

MATERIAL PROPERTIES
Conductive, Responsive, Inorganic

The Hugo Red fragrance bottle draws on references to a canteen, a water bottle used by hikers that features a screw top attached to the bottle neck with a short strip.

For the German fashion brand Hugo Boss, a certain yin/yang lies at the core of their stylish ethos. From fashion to fragrance, what is cool is also what is hot, with many of their designs channeling a quiet luxury tinged with an edgy attitude. To celebrate the twentieth anniversary of its couture label Hugo, the fragrance division launched Hugo Red. The fragrance is, appropriately, built around contrasting warm and cool notes, or accords, a blend of essences that are carefully chosen to create a new and different scent. It opens with "Solid Chill," based on grapefruit and rhubarb, and is followed by "Liquid Heat," which features cedar wood and hot amber.

While Hugo Red follows the naming structure established by the brand's color-oriented apparel lines—Boss Orange for casualwear and Boss Green for sportswear—the warm/cool concept prevails in the emblematic red flacon. Shaped like a hikers' water bottle, the exterior is coated in opaque, matte-red thermochromic ink that changes color according to the heat of the user's hand. When the wearer picks up the bottle, the warmth generates a handprint that remains on the bottle, lingering before eventually disappearing from the surface.

The effect seeks to express the fiery intensity of attraction, a key role of any fragrance, as well as the telltale sign of having been touched, a fingerprint. The result is a personal, intimate experience, evoking the moment between the feel of another's skin and its evanescence.

MATERIAL INSIGHT

The coating applied to this bottle changes color when it is heated above a specific "activation" temperature, in this case the warmth of a hand grasping it. The color change, in a process known as thermochromism, causes the surface to morph from scarlet to a different shade of red. The pigments used in this type of technology are liquid crystals, which

cause the color change. They have a periodic spacing of layers on the molecular level that can change on variation in temperature, causing light to diffract differently and thus the change in color. The process is repeatable once the coating has cooled down and returned to its scarlet color. The specific temperature of color change is engineered within the liquid crystals to ensure reliable activation.

The brand's edgy, urban attitude is reflected in a coating of matte-red thermochromic ink that changes color in response to the heat given off by a user's hand.

BALLANTINE'S FINEST "LISTEN TO YOUR BEAT" BOTTLE

DESIGNER
Andy Atkins, design director,
Hornall Anderson UK
www.hornallanderson.com

DESIGNER BIOGRAPHY
Hornall Anderson specializes in the design of branded experiences, with creative expertise that spans print, digital, packaging, and environments. With offices in Seattle, New York, and the United Kingdom, Hornall Anderson is part of the DAS Group of companies, a division of Omnicom.

CLIENT/MANUFACTURER
Pernod Ricard
www.pernod-ricard.com

MATERIAL
Electroluminescent technology

MATERIAL PROPERTIES
Electrically Charged, Printable, Programmable

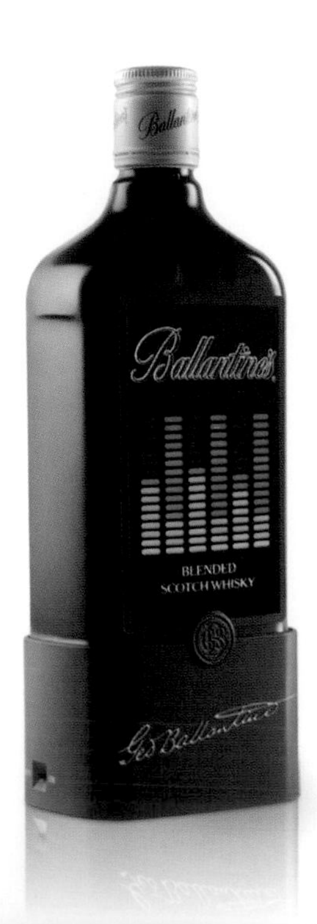

With the miniaturization of electronics, packaging takes on new capabilities. To draw attention to Ballantine's Finest in a dimly lit bar, an illuminated sequence mimics a graphic equalizer as if the bottle is responding to music.

For a liquor purveyor, the nightclub is king. But in a world where the lights are deliberately low, standing out is a challenge, especially for an inert bottle trying to capture the attention of a young clubber with plenty of competing options before him or her. Such was the scenario Hornall Anderson UK faced when searching for a way to give shelf-presence to Ballantine's Finest amber liquor.

The answer lay in a simple device: light. The design firm achieved a dazzling effect by coating bottles of Scotch whisky in a specially developed electrostatic ink that, when triggered by a battery unit housed in the base, produces the world's first self-contained electroluminescent bottle. When activated, lights rise and fall through six different illuminated sequences, mimicking a graphic equalizer, as if the bottle were responding to the musical beat in the room and an integral part of the party.

The striking results were the culmination of multiple experiments involving power sources, electrical wiring, color filters, and safety features (strapping electronics on to a liquor bottle in a crowded room generates its own concerns). And then there was the matter of how to get the finished bottles down the production line in Scotland so they could be properly packed and shipped. None of this, however, is apparent to the consumer, a hallmark of effective design. What they are left with as they amble home from a night out is the brilliance of Ballantine's.

MATERIAL INSIGHT

The use of electroluminescent (EL) inks printed onto the surface of the bottle in place of a standard label creates a dynamic interactive display that responds much like a graphic equalizer. Unlike other EL technologies printed on bottles, this version is a self-contained unit with a battery pack and control module contained within a close-fitting

shoe or skirt on the base of the bottle. The bottle itself is spray-coated with a deep-blue pigment to blend in with the base and create a more standout effect for the EL technology. Electroluminescent inks are essentially polymer thick films that respond to electricity by emitting photons (light). They tend to do this in the blue part of the spectrum, giving off a blue or green light that can be changed in color by using filters.

Electroluminescent technology allows thin materials to emit light. The illusion of light moving across the surface of a bottle produces an optical effect that speaks to design-driven consumers of premium whisky.

T+INK

DESIGNER
Touchcode
www.t-ink.com
www.touchcode.de

DESIGNER BIOGRAPHY
T+ink Inc. is an innovative technology company that utilizes proprietary "thinking inks" to enhance products through the creation of applications in the rapidly expanding printed electronics industry. It is the leading designer and manufacturer of unique and innovative conductive inks, coatings, and applications used to significantly enhance the interactive capabilities of a range of products.

MANUFACTURER
T+ink
www.touchcode.com

MATERIAL
Flexible conductive ink that can act as an electrical switch and respond to a capacitive touch surface

MATERIAL PROPERTIES
Invisible, Recyclable, Robust

There is ink as most people know it. And then there is T+ink. What a difference that single letter makes to this age-old medium. The T stands for "thinking," but it could just as well be "transforming," since that is the effect produced by embedding printed objects with invisible codes that make them smart, interactive, and secure.

Conductive ink technology allows consumers to engage with a wide range of printed media—books, magazines, advertising, tickets, and product packaging. This is done—as shown in the Moto X print advertisement—by printing buttons on a page, or in the case of Touchcode by using a smartphone or other mobile device with a touch screen. Imagine reading a review of a cookbook in a magazine where the same page links to a recipe on your smartphone; a profile on a musician that connects to a video of him in concert; or a fashion trend report that directs to a list of in-stock retailers. T+ink's power is its ability to act as a touch spring for the multitude of places your mind might like to go in response to something you have seen or done.

None of this is conceivable without ink itself, which made typography possible hundreds of years ago. Whether words are read printed on paper or moving across a screen, they are only as good as the ideas behind them.

Touchcode prints conductive ink in specific patterns underneath the visual graphics on the card, which are then "read" by the smartphone via its touch-sensitive screen. The unique patterns instruct the phone to access a given website.

MATERIAL INSIGHT

Though not strictly packaging, this print ad for Motorola's Moto X series of smartphones is a truly interactive experience, utilizing T+ink's conductive inks to connect colored print circles on the page to the image of the phone. The printed "buttons" use conductive inks based on conductive

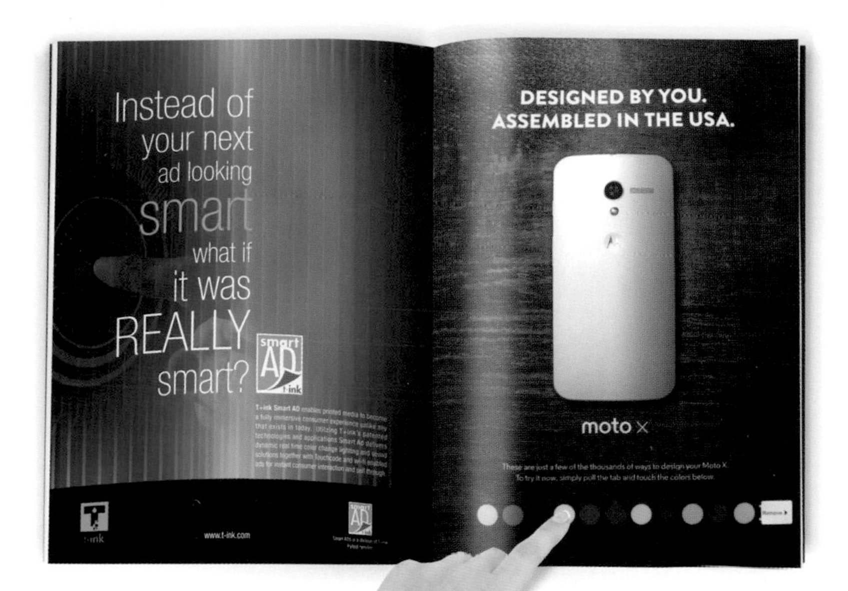

particles within a polymer ink carrier that interact with the natural conductivity of your fingers. Pressing the paper then acts as a simple switch that changes the color of the phone image to that of one of the eleven corresponding buttons. The phone image itself is in fact a simple LED membrane, powered by four watch batteries. The LED membrane is protected with a polymer film that keeps the entire composition together.

Readers of this print ad in *Wired* magazine can select the color of Motorola phone they want to see by pressing their finger on a colored circle at the bottom of the page. The conductive ink and membrane display do the rest.

SUSTAINABLE EXPANDING BOWL

DESIGNERS
Anna Glansén and Hanna Billqvist
of Tomorrow Machine
www.tomorrowmachine.se

DESIGNER BIOGRAPHIES
Product designers and co-founders Anna Glansén and Hanna Billqvist led a team who collaborated with Innventia researchers. Hjalmar Granberg works as a senior research associate on New Material and Function, Marie-Claude Béland is co-inventor, and Mikael Lindström is deputy director of packaging solutions and senior scientist in material science at Innventia.

CLIENT/MANUFACTURER
Innventia
www.innventia.com

MATERIAL
Cellulose composite

MATERIAL PROPERTIES
Organic, Malleable, Non-Toxic

Structural integrity meets dimensional intrigue in a collaboration between the Swedish creative agency Tomorrow Machine and Innventia, a research institute whose vision of the future is crafted with materials that are both renewable and smart.

Innventia's interactive paper and packaging materials, for example, change their appearance and form in response to various stimuli. True to that capability is the Sustainable Expanding Bowl, an elegant, origami-like food container that folds down flat when empty for shipping and then hardens into a vessel ideal for holding even sloshy dishes such as noodles and soups. As hot water is poured into a small hole at the top, the 100 percent biobased and biodegradable material is activated and the carton uncoils in a bit of design wizardry. As the water cools, the surface of the package hardens into a stable bowl.

An arresting study in simplicity, what begins as little more than a piece of carefully folded paper reveals itself to be an innovative vessel possessing a hallmark of great design: showing how well-made packaging solutions clearly, and without embellishment, articulate their purpose.

The use of interactive and mechano-active materials whose movements can be triggered by heat, moisture, or electricity prompts a host of creative possibilities. As Tomorrow Machine says, by combining the knowledge of scientists and the creativity of designers, it is possible to deliver "the sustainable package design of tomorrow—today."

Innventia, a leader in the development of pulp- and paper-based packaging, conceived of a 100 percent biobased and biodegradable material whose self-expanding properties would also allow for compression during transport to save space and costs.

MATERIAL INSIGHT

The term for this phenomenon of expanding when exposed to heat (in this case hot water) is "mechano-active." Materials that have this property can change their shape, rigidity, elasticity, or permeability in response to various stimuli, such as moisture, heat, or electricity. The material used for this carton is produced from a paper-based composite that uses two layers of cross-laminated sheets with differing rates and directions of expansion. These sheets strain against each other when heated and are forced to flatten out, making the carton expand. As the temperature of the water decreases, the material hardens and the bowl is set in its expanded state.

Pouring hot water into the package (**above**) triggers the mechano-active material, and a flat, folded package rises up to form an origami-like serving bowl (**below**).

GOGOL MOGOL

DESIGNER
KIAN Branding Agency
kian.ru/eng/projects/gogol_mogoleng

DESIGNER BIOGRAPHY
KIAN branding agency combines marketing ideas and innovative technological solutions. Copywriter: Kirill Konstantinov; Art Director: Evgeny Morgalev; Art Director: Mariya Sypko.

MATERIAL
Molded paper fiber, water, and calcium oxide

MATERIAL PROPERTIES
Self-Heating, Durable, Recycled, Recyclable

Rather than long, horizontal egg cartons, Gogol Mogol stacks vertically in a circular formation.

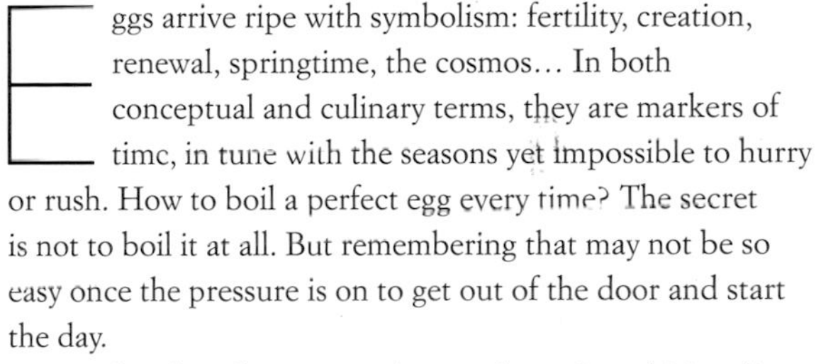

Eggs arrive ripe with symbolism: fertility, creation, renewal, springtime, the cosmos… In both conceptual and culinary terms, they are markers of time, in tune with the seasons yet impossible to hurry or rush. How to boil a perfect egg every time? The secret is not to boil it at all. But remembering that may not be so easy once the pressure is on to get out of the door and start the day.

The idea of an egg on the run drove Gogol Mogol's handheld packaging innovation. A lively green dual-purpose shell provides a convenient container for storing and transporting an egg, and can even cook the egg—faster than is possible in water. Chemicals contained in the membrane between the cardboard exterior and the egg generate heat when they are combined with water. A quick pull of the lid sets the chemical reaction in motion. Two minutes later the egg can be eaten directly from its container. Hard-boiled or soft-boiled? The choice is yours.

To give the new-to-the-marketplace design an air of familiarity, crucial for the introduction of any innovation, the Russian creative team behind the design chose a recycled paperboard for the shell akin to traditional egg cartons. Storage, transport, cooking, serving—these combined functions have the potential to change the way we interact with packaging, and even the most humble of contents.

opposite KIAN's foray into packaging innovation goes a step beyond the typical egg carton by delivering two-for-one: a protective carton that also cooks its contents perfectly every time.

printing are used to create the displays, and fine lines are possible, with the best results achieved in line widths (dot sizes) commonly used in the printing of graphics. The technology comprises a central electrolyte that controls the color, with electrochromic layers on both sides, and printed electrodes that supply the power. These are all laid onto a substrate paper and covered with a protective insulating layer that ensures the system is not affected by contact with other objects or materials.

Designed to simulate a boarding pass, an interactive ad for SATA Air urges readers to press a button and learn the promotional price for a flight. Studies show that interactive solutions boost brand recognition with consumers.

gogol
mogol
easy
egg
breakfast
pull to start

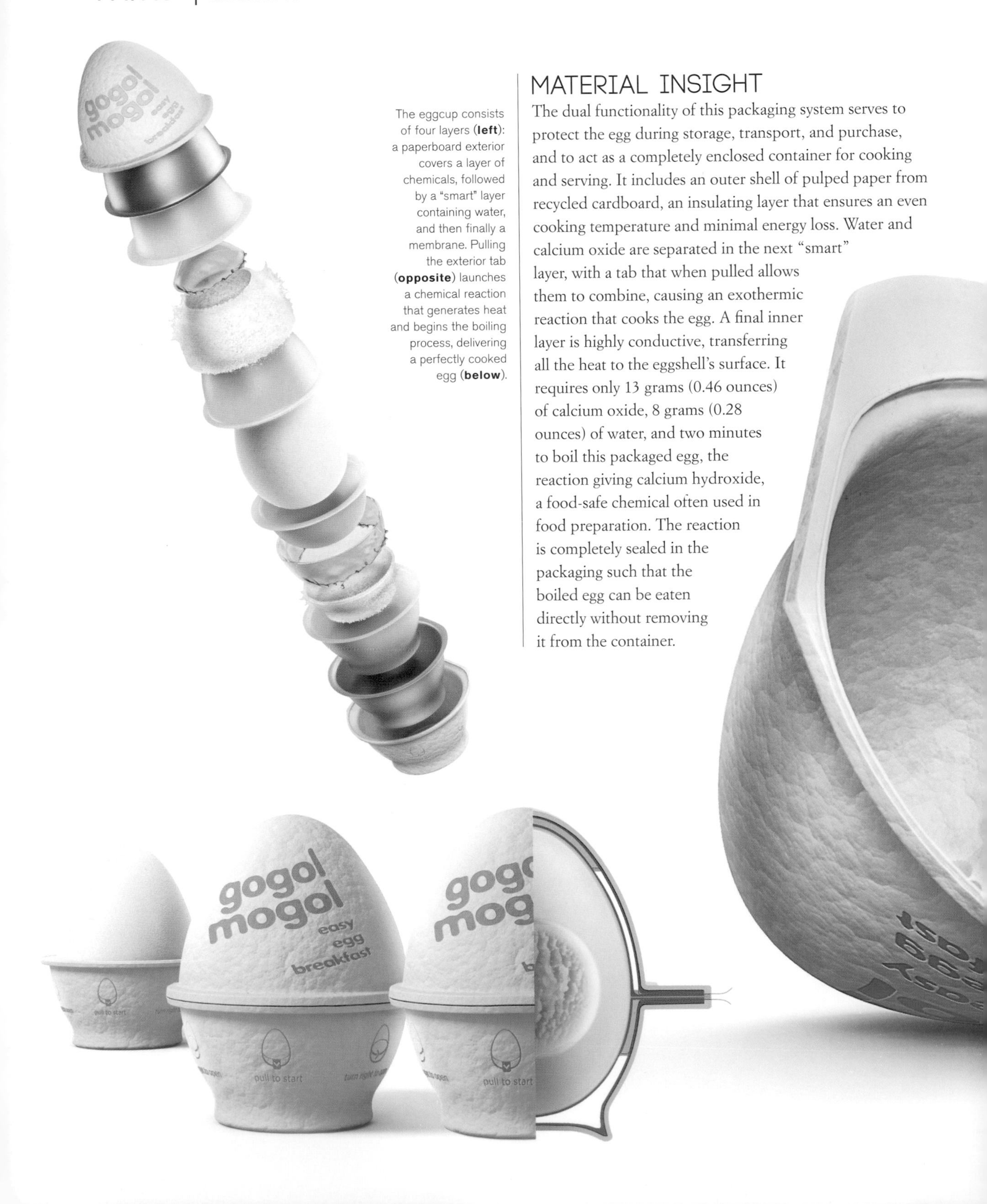

The eggcup consists of four layers (**left**): a paperboard exterior covers a layer of chemicals, followed by a "smart" layer containing water, and then finally a membrane. Pulling the exterior tab (**opposite**) launches a chemical reaction that generates heat and begins the boiling process, delivering a perfectly cooked egg (**below**).

MATERIAL INSIGHT

The dual functionality of this packaging system serves to protect the egg during storage, transport, and purchase, and to act as a completely enclosed container for cooking and serving. It includes an outer shell of pulped paper from recycled cardboard, an insulating layer that ensures an even cooking temperature and minimal energy loss. Water and calcium oxide are separated in the next "smart" layer, with a tab that when pulled allows them to combine, causing an exothermic reaction that cooks the egg. A final inner layer is highly conductive, transferring all the heat to the eggshell's surface. It requires only 13 grams (0.46 ounces) of calcium oxide, 8 grams (0.28 ounces) of water, and two minutes to boil this packaged egg, the reaction giving calcium hydroxide, a food-safe chemical often used in food preparation. The reaction is completely sealed in the packaging such that the boiled egg can be eaten directly without removing it from the container.

turn right to open
pull to start

BOMBAY SAPPHIRE ELECTRO

DESIGNER
Dominic Burke, creative director, Webb deVlam
www.webbdevlam.com

DESIGNER BIOGRAPHY
Dominic Burke has been part of the creative team at Webb deVlam since 2003. Previously a senior designer at Fitch, he is the recipient of numerous gold, silver, and bronze packaging design awards.

CLIENT
Bacardi Global Travel Retail
www.bacardilimited.com

MANUFACTURER
Karl Knauer KG
www.karlknauer.de
ROX Asia Consultancy Ltd
www.roxasia.com

MATERIAL
Electroluminescent ink powered by a battery on the base of the pack and activated by a mechanical switch

MATERIAL PROPERTIES
Animated, Surface Effects, Printable

In the maelstrom that exists today for modern travelers shuttling between international flights and time zones, duty-free shopping carries its own allure. Although neither reliably cheaper nor exclusive, these oases of luxury items, which seem to revel in fluorescent lighting and shoppers with barely a few hours' sleep, have proved their financial worth. Accordingly, brand-name distillers have taken to issuing their own unique packaging designs and retail displays specifically for the sector.

Among the standouts has been Bombay Sapphire's Electro pack, a carton adorned with an eighteen-second light show that is powered by a battery in the base of the box. Whenever the pack is picked up from the shelf, a hidden mechanical switch is triggered, illuminating the lush exuberance of artist Yehrin Tong's illustration, which in its swirling shapes mimics the distinctive vapor infusion process so integral to the gin's flavor. The maximalist bang provided by Tong's intricate patterns is well suited to the sequential cascading effect produced by an electrical current that runs through the different pathways of the design.

In keeping with the overall brand philosophy of "infused with imagination," the luminous effect visually transports the customer to unexpected places, elevating pure decor into a surprising and artful fireworks display that can be repeated at will. The intrigue produced by trying to figure out how it works explains the British distiller's leading position in travel retail.

MATERIAL INSIGHT

Electroluminescent (EL) inks emit a low-intensity glow when charged with an electric current. The inks used on the secondary (outer) packaging are screen-printed in a specific design as a final process following initial graphic printing. The inks contain the electroluminescent chemical zinc sulphide doped with copper or silver. They are connected electrically to a battery pack that charges the EL and lights the packaging. As a potential buyer picks up the packaging at the point of sale, an intuitive mechanism activates the light animation on the front of the packaging. A five-stage light sequence is shown: First the bottle image lights up, followed by luminous design elements surrounding the image. The cycle lasts a total of eighteen seconds. It then stops, restarting only when the package is moved again. After the batteries are removed, the packaging can be disposed of in the usual manner. The innovative light animations can continue to be displayed by simply replacing the batteries.

opposite Illustrator Yehrin Tong's hypnotic plumes are reproduced using electroluminescent ink, which glows when charged.

right When the package is picked up, a current housed in a battery pack on the base of the box is activated, illuminating the design over an eighteen-second cycle.

HEINEKEN STR

DESIGNER
Heineken in collaboration with dBOD
www.dbod.nl

DESIGNER BIOGRAPHIES
Mark van Iterson, global head of Design Heineken
John McGuire, project manager, Packaging Innovation Heineken
Erik Wadman, design director, dBOD
Peter Eisen, designer
Pascal Duvall, illustration, Iris

CLIENT
Heineken
www.heineken.com

MATERIAL
Aluminum and UV-sensitive ink

MATERIAL PROPERTIES
Printable, Lightweight, Conductive

below and opposite The STR bottle is fashioned from recycled aluminum printed with UV-sensitive ink that under "black light" reveals a version of the iconic star logo.

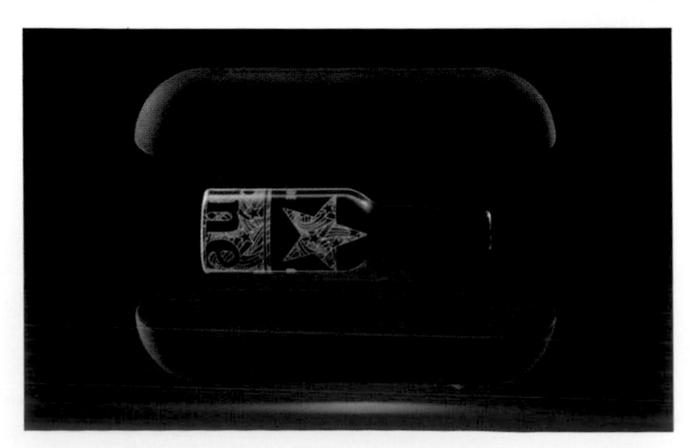

In natural light, the Heineken STR looks almost like a typical bottle from the famed Dutch brewer, even with a shimmery, cool aluminum container replacing the well-established glass bottle. Under ultraviolet light, though, the full effect of the change is revealed. UV-sensitive ink has been used to create a graphic of stars and streaks that is only visible in the dark. The recyclable bottle lights up, wrapped in a twinkling sky; beer as a stairway to heaven?

The design's luminosity—ideal for the dimly lit clubs and bars around the world where the limited-edition bottle was introduced—breathes modernity and lightheartedness into the upscale, heritage brew. Intended for design-minded consumers, the many small details that make up the classic Heineken label have been simplified, and key icons, such as the ubiquitous star, given a new prominence and purple hue.

After 140 years of careful brand management, the Heineken star is as much a part of the brand's iconography as Coca-Cola's flowing script or Starbucks' twin-tailed siren is of theirs. Repeat viewing of the symbol means that even under a black light, the shape's warm, festive, cheerful mood emanates. The choice of aluminum gives the brand a new tactile dimension: seamless, lightweight, and resilient.

MATERIAL INSIGHT

The bottle itself was printed with a bold, graphic, UV-sensitive ink, which was invisible in regular light but lit up in blue and white under the UV or "black light" typically used in nightclubs. Prior to its launch, the STR bottle was distributed in a bespoke capsule to Heineken brand managers across the globe. The all-black, 3D-printed capsule was divided horizontally and held together by a magnetic lock. When opened, a UV light was automatically activated, revealing the UV print. The 3D-printed capsule was created by Freedom of Creation using selective laser sintering, with a pillow-embossed logo on the outside that enabled it to stand.

ADE
ine

SHIGENO ARAKI

CHAPTER 6
MASS CRAFT

In its simplest form, mass craft is the piece of string tied around a high-volume packaged-condiment jar, or the wax used to cover the cap of a whiskey bottle, both of which suggest hand-crafted attention—with its inevitable blemishes—to a mass-produced item. It is the ability to create products that are differentiated from mass-customized products, such as Nike ID, Share a Coke (with individual names printed on labels), or monogrammed towels and shirts; it offers the real effects of "craft," allowing imperfections, the "hand of the maker," and the variations from piece to piece expected of individually crafted products. This ability to manufacture unique pieces has been lost in high-volume manufacturing owing to a need for efficiency, low cost, compliance, and reliability, resulting in pieces indistinguishable from one another. The advent of digital production processes such as printing, additive manufacturing, and CNC (computer numerical control) machining, however, is beginning to free us from some of these constraints, giving us greater versatility to achieve noticeable differences between any two examples of the same product, even when producing hundreds of thousands.

The Absolut Unique bottle is perhaps the most effective demonstration of both the successful results achievable and the degree of complexity required for mass craft. The design created on the surface of the glass for each of the nearly four million bottles was completely different from the next in both color and pattern. The colors were hand-mixed and selected by staff daily, then sprayed on by a machine that was controlled to create randomness using a mathematical algorithm. Direct digital printing—where

opposite Hyperform explores folding as a design strategy for compressing large-scale objects into the small volume of desktop 3D printers. This translucent polymer chandelier is produced as one long chain that, when removed from the printer, automatically forms the correct geometry. The parts bend at right-angles in predetermined ways so that no instruction booklet is needed to assemble the fixture.

A high-resolution print app offered through adidas.com allows wearers to custom-print personal photographs onto Adidas ZX Flux sneakers. Artwork sent via the app is printed onto the surface of the textile prior to fabrication.

logos, information, and patterning are printed directly onto each package rather than on a pre-formed label or sleeve—also allows this degree of variation for each individual piece and even gives greater control and precision when deciding on design.

Although not currently a commercial aspect of packaging, it is likely that the adoption of AM (additive manufacturing) methods in addition to our existing production processes should allow for greater exploration of the "craft" effect. AM has seen an incredible rise in popularity for one-off projects, but also increasingly as a method of creating unique designs in low and medium production volumes. Jewelry, sneakers, toys, furniture hardware, and construction parts are now all being produced using various AM methods, giving opportunities to create one-of-a-kind pieces that exhibit mass-craft aesthetics. The current slow speeds of production for 3D printing compared to, say, injection molding or paper-box die-cutting mean that it cannot yet quite achieve speed and cost parity with many of the existing package-production processes in high-volume applications. However, the likely hybridization of manufacturing processes over the next few years—where AM is used as a minor but very visible "add-on" to a higher-volume process—will give designers a chance to add the unique details that they are currently unable to.

When we talk about "organic" lines or effects, it is the perfect imperfection of nature that we refer to. In its simplest definition, organic design is the absence of straight lines and sharp corners, and truly "organic" forms should contain those normal variations inherent in natural systems throughout all aspects of the design. From biomimicry to the latest methods of genetic modification, we are using nature in more controlled and advantageous ways to enhance our abilities to create. The newest thinking on diet that considers everything in relation to the microorganisms in our gut is just one aspect of the science that is discovering how essential bacteria are to every part of our environment. The current crop of biopolymers, which rely on bacteria for production and experimentation, is also being formulated with an eye toward the fabrics, shoes, and even electronics that can be created by these incredible organisms.

Recent developments in bio-led design and manufacturing are demonstrated by the packaging of yoghurt snacks by Stonyfield, which uses grown membranes for protection that are themselves edible. What if we were able to use nature to create decorative surfaces, using its unique production methods to give us graphic effects, protection, water resistance, and all the other performance attributes we might want in a surface? If so, we can expect a new level of beauty from our products, one that relies on natural processes to create function on a truly nano scale (as nature inherently does), and a perfect imperfection that is universally understood and appreciated.

The question then arises in the expansion of this trend about how the wider public will take to products that have not only uniqueness but also imperfections built into them. Almost a decade ago, Material ConneXion hosted a conference titled Malfatto, an event that explored the idea of the "poorly made"—a tongue-in-cheek designation of an approach to design that privileges controlled experimentation, the hand of the maker, and materials research that is not directly related to industrial thinking. The conference was an attempt to give voice to an innate understanding that there is great value and beauty in products that have that human touch. The concept is now perhaps of even greater import, as we see the number of mass-craft products expanding.

As a design trend, Malfatto is perhaps one of the most complex, commingling as it does the impersonal nature of most modern production, the value of craft, our love of slight imperfections, and the simple human enjoyment of creating. Malfatto is related to mass craft, perhaps its unruly older brother, less concerned about pleasing but more true to itself and its worth. Care must be taken, however, with the current direction of this trend of showing the "hand of the maker," so that viable efficient production is not lost to the "artisan" vision, yet retains some of the playfulness of its sibling.

Perfume and cosmetics packaging by Technotraf is machined from solid lumber using high-precision turners and CNC routing to produce complex surfaces.

WIKIPEARLS

DESIGNER
Dr. David Edwards, designer François Azambourg, and biologist Dr. Don E. Ingber
www.lelaboratoire.org
www.wikipearls.com

DESIGNER BIOGRAPHY
A creator, writer, and Harvard professor, Dr. David Edwards is the founder of WikiFoods Inc. He is also founder and director of ArtScience Labs, a network of three cultural innovation labs with homes in Paris and Cambridge, Massachusetts. At ArtScience Labs, cultural experiments at the frontiers of science with leading artists, designers, engineers, scientists, and students produce works of art and design that are exhibited in educational, cultural, and commercial galleries.

MATERIAL
Chocolate, dried fruit, nuts, seeds, and other natural substances, and chitosan

MATERIAL PROPERTIES
Edible, Biodegradable, Renewable

opposite WikiPearl mimics the way fruits and vegetables employ their own protective skin, shielding edible contents from water loss and contamination.

If "grab-and-go" dining counts among the benefits of modern life, then vast numbers of us are headed for an encounter with WikiPearls, bite-sized, non-plastic packaging that exemplifies the concept of biomimicry—using nature's designs and processes to solve human problems. Much like a grape, or any naturally self-protective food, these portable, edible "pearls" involve two basic elements—a heart and a skin. "Hearts" ranging from soup, ice cream, and yogurt to fruits and vegetables are encased in natural, biodegradable "skins," which are safe to eat yet impervious to water and oxygen.

In the ongoing quest to minimize packaging, WikiPearls offer a way forward, marrying advances in culinary technology with the sensorial experience of food. As a viable packaging solution, they go beyond mere matters of taste to serve as an efficient conveyor of perishable ingredients with an added nutritional boost.

Cross-disciplinary exchange able to spark new ideas has become almost commonplace, and WikiPearls are no exception. The idea was ignited during an exchange between Dr. David Edwards, a Harvard professor, and Ken Snelson, a New York sculptor. Inspired by the possibilities, Edwards then embarked on a longer exploration with Harvard students around transporting water in ways akin to biological cells. He next collaborated with designer François Azambourg and biologist Don Ingber, whose findings were made public in Paris in 2010, followed by the founding of WikiFoods in Cambridge, Massachusetts, in 2012. While evoking the momentary indulgence of a bon-bon, WikiPearls are nonetheless an emphatic statement on the possibilities of twenty-first-century food packaging.

Bite-sized portions encapsulated in a protective, cell-like membrane combine the primal attraction of food with the gastronomic science of electrostatically charged food, a biopolymer, and ion particles.

MATERIAL INSIGHT

This packaging concept acts as a water and oxygen barrier—keeping the correct amount of moisture within the product while protecting against the ingress of contaminants. The WikiPearl skin itself is an "electrostatic gel that has been created by harnessing interactions between natural food particles, such as fruit, nuts, and coconut flakes, along with small amounts of fibrous chitosan, algae extract and calcium ions," with a tensile strength sufficient to hold its contents securely. The intention is to consume the skin with the food inside, in this case yoghurt (though ice cream, cheese, vegetables, even cocktails are being considered), flavoring the "packaging" to enhance the consumption experience.

ABSOLUT UNIQUE

DESIGNER
John Lagerqvist, creative director and partner, and Mårten Knutsson, partner, Family Business
www.familybusiness.se

DESIGNER BIOGRAPHY
John Lagerqvist is creative director and co-owner of Family Business, a Stockholm-based brand activation agency renowned for its work for Absolut Vodka and other global clients. The agency is the recipient of gold awards from The One Show, Cannes Lions, Eurobest, FAB Awards and D&AD, among others.

CLIENT
The Absolut Company, a Pernod Ricard brand
www.absolut.com

MANUFACTURER
Ardagh Group
www.ardaghgroup.com

MATERIAL
Glass and ink

MATERIAL PROPERTIES
Printable, Recyclable, One of a Kind

The relationship between culture and commerce is constantly evolving. With Absolut Unique, the Swedish vodka purveyor references Abstract Expressionism to evoke a festive, celebratory mood. For this annual limited holiday edition, splashy colors and the "physical immediacy" of ink, to borrow a term applied to Abstract Expressionism, are on full display across the surface of four million bottles, comprising an unusually large set of unique, collectible packaging designs. The kaleidoscopic results mimic spin paintings, works of art both naive and sophisticated that are made by rotating a canvas on a flat surface and applying unrestrained color to achieve a dramatic, even hypnotic effect.

For the bottles, a combination of printing techniques was needed to ensure the production of so many uniquely designed bottles. First the bottles were spray-coated with three different colors via a computerized valve and control system, after which they were burned. Thereafter different patterns/artworks were printed on the bottles. In total, thirty-two coating colors were used and fifty-one different patterns. Lastly, numbered paper labels were applied. To verify the "uniqueness" of so many bottles, Absolut engaged a quality-management professor to calculate the precise range of variations and guarantee the results.

Once the bottles were ready for shipping, they were randomly mixed between cartons so that they arrived at the distributor in a non-sequential combination, allowing for yet one more element of chance to enter the equation.

MATERIAL INSIGHT

The use of cutting-edge spray-coating technology to pattern the outer surface of the glass gives each of the close to four million bottles a unique design. To achieve this, every day a new batch of paint colors was hand-selected by the design

team and fed into spray consoles that were controlled using an algorithm, creating "carefully orchestrated randomness" to ensure each spray pattern was different from any other. This combination of color and patterning through computation means that ninety-four quintillion (10^{18}) uniquely designed bottles could be produced before two identical ones appeared.

When no two bottle designs are exactly alike, each product becomes a coveted, exclusive object, desired as much for its artistic expression as for its contents.

ABSOLUT ORIGINALITY

DESIGNER
Magnus Tear, creative director, happytear
www.happytear.com

DESIGNER BIOGRAPHY
Magnus Tear is CEO and creative director of happytear, a Stockholm-based creative agency whose expertise is in film, fashion, design, and retail.

CLIENT
The Absolut Company, a Pernod Ricard brand
www.absolut.com

MANUFACTURER
Ardagh Group
www.ardaghgroup.com

MATERIAL
Clear and cobalt glass

MATERIAL PROPERTIES
Rigid, Second Life, Transparent

Each year, with the advent of the winter holidays, Absolut embarks on one of its now infamous creative endeavors, bringing fresh inspiration to its iconic bottle. In recent years the glass container, a shape based on an eighteenth-century apothecary design, has become its own canvas, a surface ripe for experimentation.

For Absolut Originality, the concept was simple—produce a one-of-a-kind, graphically compelling visual solution. To achieve this, the Swedish company looked to its own origins, taking inspiration from the country's outstanding heritage of glass innovation by infusing the material with color—a single dollop of cobalt is inserted into each mold while the glass is still hot, threading its way down the entire length of the bottle. Evoking the simple, even ethereal aesthetic that is intrinsic to Scandinavia, each indigo-streaked vessel conveys Absolut's adherence to design clarity over decorative embellishments. Visions of clear, tumbling streams amid thick forests and the cool purity of a world blanketed in snow seem to tumble from every rim.

MATERIAL INSIGHT

Building on the success of Absolut Unique, the Originality line used a small amount of cobalt blue diffused into the glass used to create each bottle. Cobalt blue is a pigment produced using cobalt salts of alumina. The compound is made by sintering finely ground CoO (cobalt oxide) and Al_2O_3 (alumina) at 1,200°C (2,192°F). When added to the molten clear glass at 1,100°C (2,012°F) as it enters the mold, it streams down the inside of the bottle, creating a unique streak of blue. At that temperature the cobalt is invisible, but as the glass cools off, a beautiful and unique blue infusion appears, with each of the four million bottles an original. Though not as strikingly "unique" as its predecessor, it requires no additional step in the manufacturing to create the effect and is also reminiscent of artisanal blown glass.

left and opposite Inspired by Sweden's traditional glass craft, a single dollop of cobalt glass introduced at random contrasts with the otherwise crystal-clear bottle and results in four million one-of-a-kind pieces of art.

ABSOLUT
Country of Sweden
VODKA
Enjoy this limited collection of true originality, ABSOLUT VODKA in 4 million designs.
No. 3 076 547/4 000 000
40% ALC./VOL. 750 ML
SWEDISH VODKA
ABSOLUT SINCE 1879
ABSOLUT
Country of Sweden
VODKA
Enjoy this limited collection of true originality, ABSOLUT VODKA in 4 million designs.
No. 3 147 914/4 000 000
40% ALC./VOL. 750 ML

MARMITE XO

DESIGNER
Matt Gandy, design director,
Hornall Anderson UK
www.hornallanderson.com

DESIGNER BIOGRAPHY
Hornall Anderson specializes in the design of branded experiences, with creative expertise that spans print, digital, packaging, and environments. With offices in Seattle, New York, and the United Kingdom, Hornall Anderson is part of the DAS Group of companies, a division of Omnicom.

MANUFACTURER
Unilever
www.unilever.com

CLIENT
Marmite

MATERIAL
Glass, wax, and fabric ribbon.

MATERIAL PROPERTIES
Non-Toxic, Recyclable, One of a Kind

left and opposite
Taking full advantage of the creative possibilities afforded by a collectible, limited-edition design, the Marmite XO packaging possesses numerous handcrafted details including a gold wax seal stamped with a crest.

The handcrafted care extended to the packaging of Marmite XO, an extra-mature variant of the infamous yeast extract, epitomizes the passion felt by the most ardent of the sticky food paste's fans. Lovingly wrapped and sealed in its signature dark-brown jar that mimics a "*marmite*" (French for a large covered earthenware or metal cooking pot), the XO version alludes to both the extra-strong nature of the contents and the merits of an accompanying social media campaign that invited hardcore consumers to express their enthusiasm for the brand. Select participants in the campaign were anointed "the Marmarati," a mysterious, exclusive inner circle of Marmite Lovers whose role was to help develop the new Marmite flavor, champion the limited-edition run of XO, and convert new brand advocates.

Tasked with creating a bespoke, collectible piece to reward the Marmarati for their devotion and commitment, Hornall Anderson UK set up a limited-edition production line in their offices. Inspired by luxury wines and spirits, the overall design is rich with vintage cues. Each glass jar was sealed with hot gold wax and stamped with a heraldic crest designed especially for the Marmarati. The lids were finished with a watchstrap and individually signed and numbered.

A few singed fingers later, the Marmite Food Company's Victorian origins seemed to have come full circle, only now with an established following far beyond the yeast paste's original, accidental invention.

MATERIAL INSIGHT

Part of a larger, extended roll-out marketing campaign, the packaging for this more intense and stronger-tasting Marmite demonstrates some of the simple additions that can effectively present a mass craft effect. Using wax for a gold seal on the back face of the glass jar as well as a black-pigmented sealant to cover the cap meant that each one

of this limited-edition XO ("extra old") series was slightly different. In addition, the number of each individual piece of the total 200 was handwritten and signed in script on the "watchstrap" black ribbon that covered the cap prior to wax sealing. Although in this case these packaging details were produced by hand, the process lends itself to more high-volume production using existing manufacturing techniques.

FRUIT BOX

DESIGNER
mcgarrybowen São Paulo (ex-AGE Isobar); creative director: Carlos Domingos; art direction: Henrique Mattos, Cristiano Rodrigues, Fuku
www.isobar.com

DESIGNER BIOGRAPHY
Carlos Domingos is chief creative officer of mcgarrybowen São Paulo (ex-AGE Isobar). A respected leader in the Brazilian advertising industry, he is the copywriter behind numerous memorable campaigns. He was the founder, president, and chief creative officer of Age Isobar, and previously worked for J. Walter Thompson and W/Brasil, and was creative director of DM9DDB when it was elected Agency of the Year at the Cannes Festival for two consecutive years (1997–98).

MANUFACTURER
General Brands
www.generalbrands.com.br

CLIENT
Camp Nectar

MATERIAL
Fruit

MATERIAL PROPERTIES
Biomimicry, Malleable, Perishable

During a two-year experiment intended to give authenticity to "Made from Real Fruit," the Brazilian juice company Camp Nectar cultivated fruit shaped to resemble juice boxes.

Camp Nectar, with its suggestion of lavish greenery all for the playful pleasure of bees, nonetheless possesses a utilitarian ethos. To showcase the 100 percent pure ingredients in their all-natural Camp Juice, General Brands in Brazil asked the advertising agency mcgarrybowen São Paulo for a fresh take on their "Made from Real Fruit" tagline. Its ingenious yet direct response was to turn traditional juice packaging inside out by growing fruit in the shape of a box.

Camp Nectar's two-year horticultural experiment in growing specially designed fruit, complete with brand imprint, began with the clamping of clear plastic cases able to absorb sunlight over fruit buds still hanging on trees, then allowed the fruit to ripen. The process ultimately yielded 1,123 pieces of fruit in four flavors—orange, lemon, guava, and passion fruit—all in the shape of a Camp Nectar juice box. The products were then displayed in supermarkets right alongside other fruit.

Many companies understand the importance of innovation, but fall short when it comes to execution. While some might view the Camp Nectar solution as more gimmick than marketing genius, there is no denying the impressive commitment and investment from both agency and client to surrender to nature's timetable and see their idea through. Beyond injecting the produce section of supermarkets with a touch of modern pastorale, the campaign was widely feted at the Cannes Lions International Festival of Creativity, winning several awards.

MATERIAL INSIGHT

As an effective way to remind consumers that their products contain "all natural" ingredients, Camp Nectar grew fruit in the shape of their product. The company created plastic molds in the shape of the juice boxes, then wrapped them around budding lemons, oranges, guava, and passion fruit.

It took two years of research to achieve the right process. These molds were affixed around the growing fruit at the early stages, and removed only once the mold was filled. The 1,123 box-shaped pieces of fruit were displayed in supermarkets to tout the fact that Camp Nectar's juice is full of real fruit. Though not for sale, the fruit is in fact edible and is not harmed by the shape and space constraints of the transparent, thermoformed plastic molds.

Customized molds were placed around budding oranges, lemons, passion fruits, and guavas so that as they grew they assumed the shape of their cartons, including a straw and flaps.

PANTONE BEER

DESIGNER
Txaber Mentxaka
www.txaber.net

DESIGNER BIOGRAPHY
Born in 1969 in Bilbao, Spain, Txaber studied Graphic Design at CVNP (Bilbao) and since then has worked for various design studios. He also pursues his own independent projects, many of which can be seen on Behance (behance.net).

MANUFACTURER
Protoype Design

MATERIAL
Aluminum and ink

MATERIAL PROPERTIES
Shelf Impact, Simplification

If the growing consensus is that we are in an era of design-based thinking increasingly applied to mainstream business, then there is something inevitable about a line of beer packaging that communicates through color. Inspired by the Pantone color-matching system, Spanish designer Txaber had the idea to match the shade of the beer inside with its corresponding numerical shade. Lined up sequentially, the tonal range progresses from shimmering citron through tawny golds and ambers to umbers and brown-blacks. The finishing touch? The font Hipstelvetica Bold by Spanish typographer José Gomes. "At first I chose Helvetica, but the typeface from José is more personal and modern," Txaber explains.

Now that Nespresso has educated consumers on the merits of color-coded coffee flavors, the leap to a line of beers awash in an autumnal glow is a short one. While microbreweries get the credit for breaking up the flavor monopoly of national brands and spawning the craft beer movement, along with a robust audience of connoisseurs, the Nespresso pods have been a global marketing phenomenon,

turning consumers into their own baristas. Although most beer drinkers are not nearly as convenience-driven as coffee consumers, color-coded systems as a whole make life easier, neater, and more visually appealing with less mental effort. For most of us, that trumps all else. Plus, just imagine how great these would look in an ice-filled cooler.

A spectrum of color conveys the enticing diversity of beer flavors to be found, ranging from the classic to the eccentric.

MATERIAL INSIGHT

This packaging concept, which matches the color of the outside of the can or bottle to the hue of the type of beer inside, utilizes the potential wide-scale application of direct printing. Using this type of printing for containers compared to the limited print area of adhesive labels allows for a much greater canvas and the potential to cover the entire surface. The popularity of craft beers has meant that there is a greater interest in design-led presentation. This is not the first time beer has been paired with color codes: "Beertone" cards (beertone.me) contain the RGB, CMYK, and HTML (web) color codes of 202 Swiss beers, plus a set for more than 200 Brazilian craft beers.

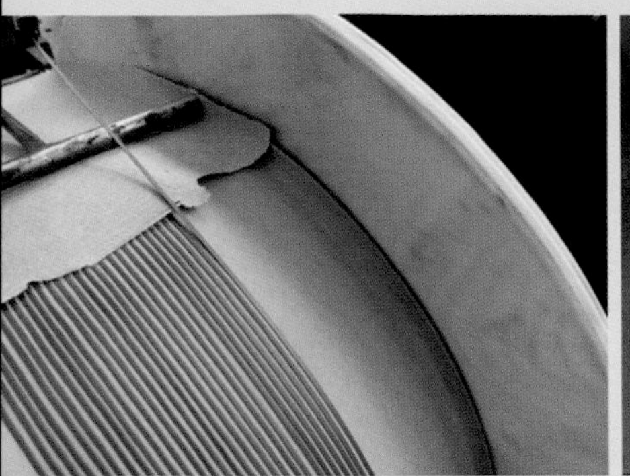

HORIZONTAL OK
VERTICAL NOT OK

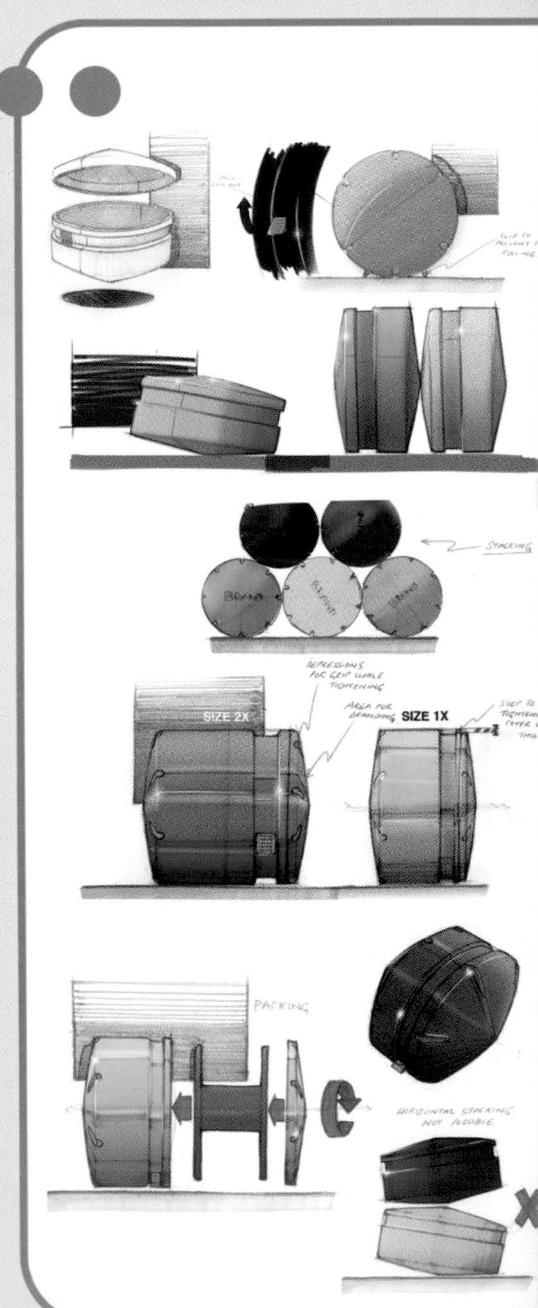
STACKING
BRAND
BRAND
BRAND
DEPRESSIONS FOR GRIP WHILE TIGHTENING
AREA FOR BRANDING
SIZE 2X
SIZE 1X
PACKING
HORIZONTAL STACKING NOT POSSIBLE

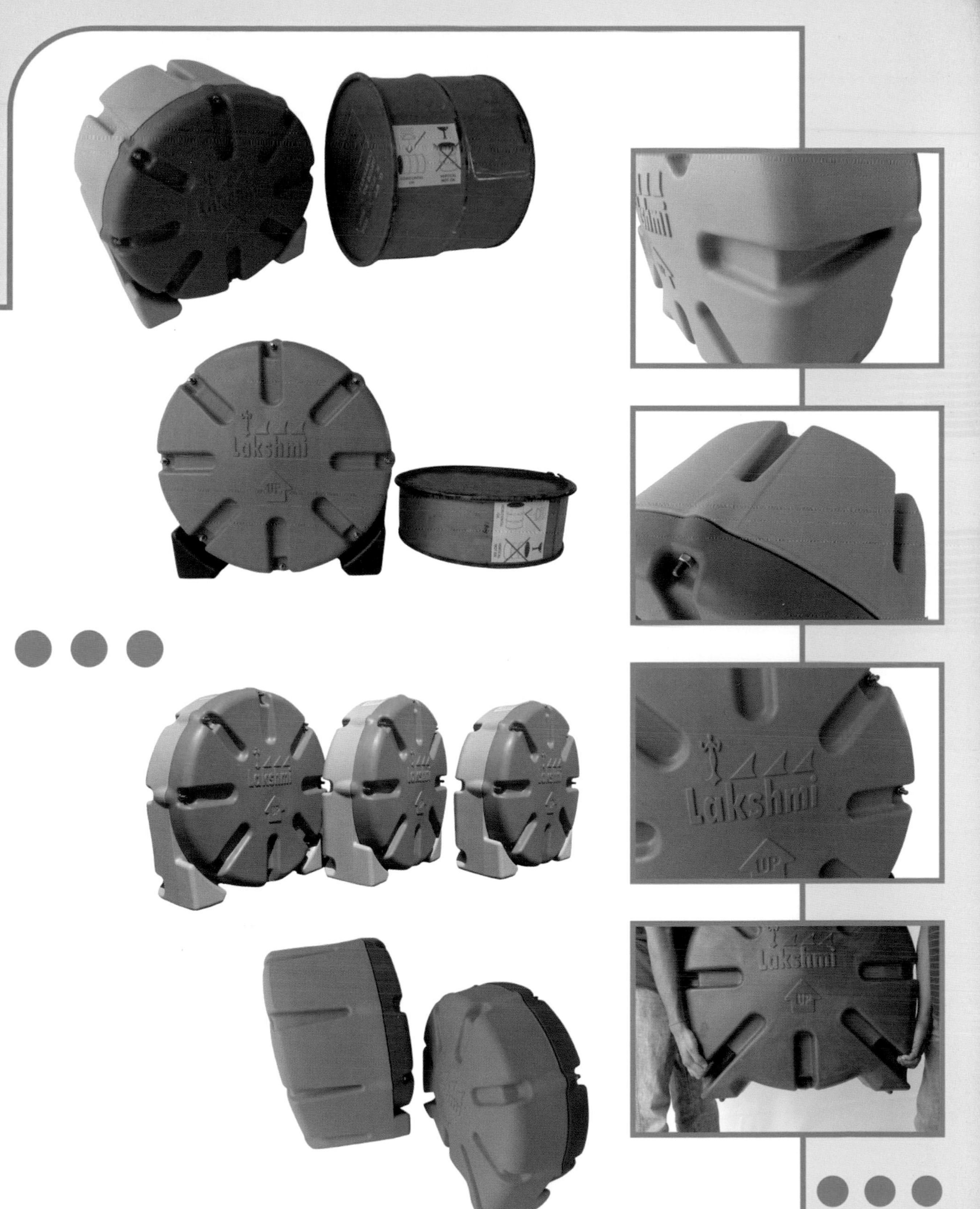
Lakshmi
UP

ORGANIC VALLEY
MILK
Welch's

MATERIAL SORTING AND RECYCLING

It takes an enlightened mind to appreciate the beauty of well-sorted trash. And yet that is exactly what is required of today's most ardent sustainable-packaging designers, as well as anyone concerned with helping to assure the future of the planet. Giant bales of crushed and bound metal cans, plastic jugs, newspapers or glass appeal not only to the orderly-minded but to anyone who dreams of a waste-free future. Sounds far-fetched? We may no longer have a choice.

A bracing report published by The World Bank—*What a Waste: A Global Review of Solid Waste Management*—interprets hard-to-gather data with the prediction that "As the world hurtles toward its urban future, the amount of municipal solid waste (MSW), one of the most important by-products of an urban lifestyle, is growing even faster than the rate of urbanization." The implication? Global waste is on pace to triple by 2100. The trash humans generate is rising fast and will not abate without "transformational" change.

It is an urgent concern that layers both pressure and broad possibility onto the role of any packaging designer—and makes a trip to a local sorting or recycling facility not just relevant but revelatory. Whether you are a sole practitioner such as Diane Bisson or Tomorrow Machine, a disruptive force seeking to alter established norms like Aidpod Yamoyo or Ecovative, or a global engine that recognizes that with scale comes the power to be a model for rippling change, such as Absolut or Virgin Atlantic, the message is the same. Recyclable, sustainable packaging possesses an inherent life cycle that can be viewed either as a short, finite line leading straight to landfill or as a circular path that is renewable and leads to a reduction in greenhouse-gas emissions. Either route depends on a shift in mindset that goes well beyond what happens at the point of purchase. The moment when a package is no longer useful and is headed for the trash demands as much consideration by designers as the moment when it is first encountered by the consumer.

"If the goal is a realistic sustainable future," writes Adam Minter in his book *Junkyard Planet*, "then it's necessary to take a look at what we can do to lengthen the lives of the products we're going to buy anyway." While ultimately sustainability is most effective when it becomes part of an organization's values and mission, packaging designers and their teams have a decisive role to play by addressing environmental issues in their projects and integrating strategies for how to reduce, reuse, and recycle into their work. Incorporating a product's life cycle into the design process, especially its recycling potential, is a process that begins, fundamentally, with a knowledge of materials.

	PLASTICS	HDPE 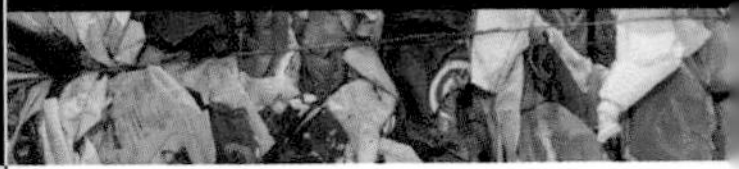
TYPES OF PACKAGING ACCEPTED AND POTENTIAL FUTURE USE		**Accepted:** Milk, water, and juice bottles, bleach and detergent bottles, margarine tubs, shampoo bottles **Recycled into:** Wood–plastic composites, outdoor decking, outdoor furniture, architectural and construction paneling, shipping pallets
SORTING METHOD	Plastics can be sorted by hand (clear water and beverage bottles are almost all PET) or by numerous other processes that differentiate by density (HDPE and PP both float, other packaging plastics sink) or by the way their chemistry reacts to electromagnetic beams such as infrared and X-ray.	Near infrared, X-ray, float/sink
RECYCLING METHOD	For all packaging plastics, mechanical recycling is the predominant method. The packages are washed and shredded into flakes, and these shredded fragments then undergo processes to eliminate impurities such as paper labels. This material is melted and often extruded into the form of pellets, which are then used to manufacture other products. Chemical recycling is possible, where the plastic is dissolved and the polymer chains are chemically or thermally pulled apart, to be repolymerized into a new "virgin" plastic.	Mechanical or chemical
HOW MANY TIMES CAN IT BE RECYCLED?		One to three times if mechanically recycled (the polymer chains are cut up during the shredding and lose performance). With chemical recycling, this does not happen, so the polymer can be recycled many more times.
CURRENT RECYCLING RATES		28% (USA)
PERCENTAGE THAT CAN GO BACK INTO PACKAGING		Up to 100%

PP	PET	PVC
Accepted: Butter and margarine tubs, yogurt containers, screw-on caps, drinking straws **Recycled into:** Durable food service items (bowls, cups, serving implements), toothbrushes, razors, reusable grocery bags	**Accepted:** Carbonated beverage bottles, water bottles, cooking-oil bottles, peanut butter jars **Recycled into:** Beverage bottles, yarn for clothing, upholstery fabrics, felt wall panels and automotive liners, reusable grocery bags	**Accepted:** Window-cleaner bottles, cooking-oil containers, detergent-powder containers **Recycled into:** Large-format banners, wood–plastic composites (outdoor decking), floor tiles
Near infrared, X-ray, float/sink	Near infrared, X-ray, float/sink	Near infrared, X-ray, float/sink
Mechanical or chemical	Mechanical or chemical	Mechanical or chemical
One to three times if mechanically recycled (the polymer chains are cut up during the shredding and lose performance). With chemical recycling, this does not happen, so the polymer can be recycled many more times.	One to three times if mechanically recycled (the polymer chains are cut up during the shredding and lose performance). With chemical recycling, this does not happen, so the polymer can be recycled many more times.	One to three times if mechanically recycled (the polymer chains are cut up during the shredding and lose performance). With chemical recycling, this does not happen, so the polymer can be recycled many more times.
Less than 1%	31% (USA) 33% (EU)	Less than 1%
Up to 50% but it is not currently used	Up to 100%	Negligible

	GLASS	ALUMINUM	STEEL
TYPES OF PACKAGING ACCEPTED AND POTENTIAL FUTURE USE	**Accepted:** Preserve and other food jars, beer, wine, and liquor bottles **Recycled into:** Solid surfacing/terrazzo, wall tiles, fiberglass insulation, food and beverage packaging, aggregate for concrete	**Accepted:** Beverage cans, aluminum foil **Recycled into:** Almost all products currently made from aluminum	**Accepted:** Beverage cans, food cans **Recycled into:** Almost all products currently made from steel
SORTING METHOD	Visual inspection (for clear, green, and brown glass) or optical sorting machine	Eddy current separator	By magnet. Most stainless steels are not magnetic, but are not often used in packaging
RECYCLING METHOD	Crushed into cullet, and remelted to form into packaging or other products	Mechanical breakdown, followed by smelting (melting and using a reducing agent to drive off impurities)	Mechanical breakdown, followed by smelting (melting and using a reducing agent to drive off impurities)
HOW MANY TIMES CAN IT BE RECYCLED?	Infinitely	Infinitely	Infinitely
CURRENT RECYCLING RATES	34% (USA) 70% (EU)	54% (USA) 67% (EU)	70% (USA) 74% (EU)
PERCENTAGE THAT CAN GO BACK INTO PACKAGING	Up to 90% (short supply makes this rare)	Up to 90%	60–80%

PAPER	OLD CORRUGATED CONTAINERS
Accepted: Mixed paper (discarded mail, telephone books, paperboard, magazines, catalogs), newsprint, high-grade de-inked paper (letterhead, copier paper, envelopes, printer scrap) **Recyled into:** Paperboard and tissue, or raw material in gypsum wallboard, roofing felt, cellulose insulation, and molded-pulp products such as egg cartons	**Accepted:** Old corrugated containers, corrugated cardboard **Recyled into:** Shipping boxes, cereal boxes, shoe boxes
By hand, or by using machines that can determine levels of fluorescence (higher-lignin-content paper—a lower-quality paper—fluoresces more brightly), stiffness, gloss level, and color	Hand sort
Used paper is mixed with water and chemicals to break it down. It is then chopped up and heated, which breaks it down further into strands of cellulose, a type of organic plant material; this resulting mixture is called pulp, or slurry. It is strained through screens, which remove any glue or plastic that may still be in the mixture, then cleaned, de-inked, bleached, and mixed with water. Then it can be made into new recycled-paper products.	Similar in process to other paper recycling. The OCC is pulped and blended with additional pulp from wood chips. Broken old fibers are blended with the new pulp to make the cardboard medium prior to blending with new pulp to make the sheet.
One to five times depending on the quality of the paper	One to five times
70% (USA) 72% (EU)	Up to 95%
Up to 100%	Up to 100%

Packaging
PRESS

MATERIALS DIRECTORY

Material ConneXion is the world's largest library of advanced innovative materials. Our online database gives users access to images, technical descriptions, and usage characteristics, as well as to manufacturer and distributor contact information, all of which has been written and compiled by our international team of material specialists. The content is intended to meet the needs of engineers and scientists as well as architects and designers by providing advanced material expertise in an accessible, user-friendly format. Each material in the database is catalogued with a six-digit MC number: the first four digits identify the manufacturer and the last two indicate how many unique materials are included in the library. Each entry has been juried by a panel of material specialists, architects, designers, and technicians from diverse manufacturing backgrounds. The jury determines whether or not a material warrants inclusion in the library by assessing its inherent innovative qualities. Material innovation is not limited to new materials and technologies. It also includes significant improvements in performance that pave the way for future development, and materials previously used in specialized fields that are becoming more accessible to designers. Innovation also applies to sustainable materials that perform as well as commonly used products with a greatly reduced impact on the environment. To learn more, visit: www.materialconnexion.com/books.

MC# 5241-02
PakNatural™; Loose Fill
Sealed Air Corporation
www.sealedair.com

Biodegradable, water-soluble, lightweight packing material made from non-food renewable materials that can serve as a sustainable substitute for Styrofoam packaging. These loose-fill packing foams are composed of modified wheat starch and are produced continuously using water as a foaming agent. The foams are 40% more dense than other loose fill such as cornstarch "peanuts." They are static dissipative and therefore do not cling to the material they surround. **Applications** include low-cost packaging for electronics, glass, and other fragile materials.

MC# 5357-02
Terra PET™
Klöckner Pentaplast GmbH & Co. KG
www.kpfilms.com

A transparent, semi-rigid polymer film that is suitable for packaging applications, containing up to 30% renewable resources. The film is produced by extrusion from amorphous polyethylene terephthalate (APET) that uses ethanol derived from sugarcane as a raw material in its production. The ethanol is converted into monoethylene glycol (MEG), which along with purified terephthalic acid (PTA) makes up the PET. Though produced from renewable resources, the performance of the film does not differ from that of standard APET. **Applications** include packaging forms such as blisters, clamshells, and trays, as well as other thermoformed applications.

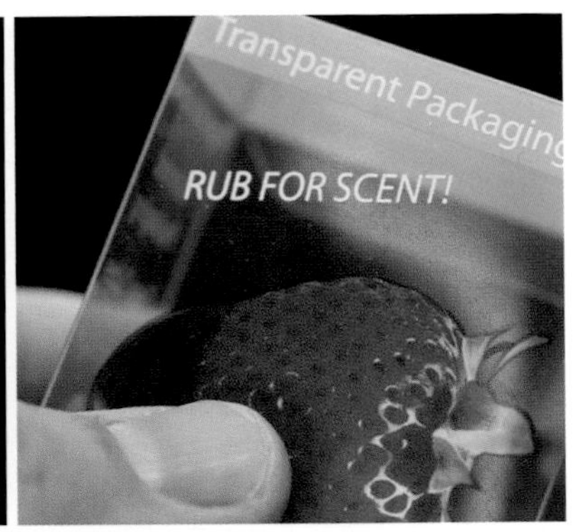

MC# 5357-03
Pentaform Smartcycle®
Klöckner Pentaplast GmbH & Co. KG
www.kpfilms.com

A clear, semi-rigid packaging film that contains post-consumer recycled (PCR) content. The film is produced using polyethylene terephthalate (PET) from both virgin sources and recycled beverage bottles. The recycled content is sourced from the company's own beverage bottle recycling facility (it had recycled more than 11.6 billion bottles as of September 2013). **Applications** include packaging forms such as blisters, clamshells, and trays, as well as other thermoformed applications.

MC# 5429-20
Keaykolour
Arjowiggins Creative Papers
www.curiousstory.com

A paper that uses an "upcycling" process to minimize the waste of natural resources. This process incorporates waste fibers that would otherwise be thrown away and reduces the use of dyes and pulp by up to 50%. The paper is FSC-certified and suitable for screen ruling, laser cutting, and hot-foil blocking printing techniques. The collection includes twenty-nine colors inspired by colors in nature. Customization is available on request. **Applications** include packaging and stationery.

MC# 5429-22
Curious Matter
Arjowiggins Creative Papers
www.curiousstory.com

Paper derived from starch extracted from potato waste, a by-product of the food industry. The paper is FSC-certified and suitable for printing and finishing techniques, lithography, letterpress, embossing, debossing, silk-screening, die stamping, and foil blocking. The paper receives renewable energy credits (RECs) from Green-e-certified energy sources to offset 100% of the electricity used to manufacture the products. It is available in seven colors: Desiree Red, Adiron Blue, Purple Majesty, Black Truffle, Andina Grey, Ibizenca Sand, and Goya White. **Applications** include brochures, invitations, packaging, and stationery.

MC# 5479-03
Plantable Packaging
UFP Technologies
www.ufpt.com

Packaging that can be "planted" after use. A variety of herb, vegetable, and flower seeds are embedded into a packaging material comprised of 100% post-consumer recycled fiberboard that is 100% recyclable. Customizable shapes and sizes are possible, including clamshells for soap packaging, hang tags for retail goods, envelopes for gift cards, or gift boxes for wine-bottle storage. **Applications** include packaging for consumer products.

MC# 5826-01
Mirel™ Natural Plastic
Metabolix
www.metabolix.com

Polyhydroxyalkanoate (PHA) polymers produced through the fermentation of natural sugars and oils. These materials range in properties, from stiff thermoplastics to highly elastic grades and grades suitable for adhesives and coatings. They may be processed using injection molding, cast film, cast sheet for thermoforming, and melt-extruded paper and board coatings. The molded material is dishwasher safe; tests have been conducted for twenty washings at 75°C (167°F). The Biodegradable Products Institute (BPI) has certified this bioplastic compostable. **Applications** include packaging, housewares, consumer durables, appliances, and disposable items.

MC# 6556-03
Mushroom® Material (Smooth)
Ecovative Design LLC
www.ecovativedesign.com
www.mushroompackaging.com

Foam-like material comprised of agricultural by-products and mushroom roots called mycelium. It is 100% renewable, has low embodied energy, and is home compostable. It starts with non-edible crop wastes such as plant stalks and seed husks. Acting like a natural, self-assembling polymer, mycelium binds the substrates together in five to seven days of growth, independent of lighting conditions, with no watering and no petrochemical inputs. The final drying procedure ensures that the materials are not alive and will not degrade until composted. **Applications** are for protective packaging, molded consumer products, and architecture.

MC# 6776-01
Plum Seed Charcoal Paper
Yamamoto Paper Inc.
www.yama-kami.com

Plum-seed charcoal paper from recycled pulp and umeboshi plum-seed charcoal, which has natural deodorizing and moisture-absorbing properties. The raw material used in the paper is plum seed that has been recovered from the plum-processing industry, a by-product that is typically thrown into landfills after food manufacturing. The seeds are ground, heated, and pressurized into charcoal and combined with pulp and water without the use of glue or resin. The resulting paper absorbs chemical vapors such as formaldehyde, acetic acid, and ammonia. **Applications** range from shoe inserts to refrigerator-shelf liners.

MC# 6782-05
Giraffe Poo Paper
G-Create (Thailand) Ltd
www.g-create.com

Handmade paper composed of 30% giraffe droppings, 40% mulberry fiber, and 30% recycled cellulose fibers. The droppings are collected from Chiangmai Night Safari and then left outdoors to dry under the sun. After they dry, the droppings are cleaned and moved to a sterilizing process, where the cellulose fiber is blended with mulberry bark and recycled paper to make a paper pulp. This paper is biodegradable, compostable, non-toxic, and acid-free. **Applications** currently include place mats, packaging (boxes, envelopes), notebooks, gift bags, and stationery.

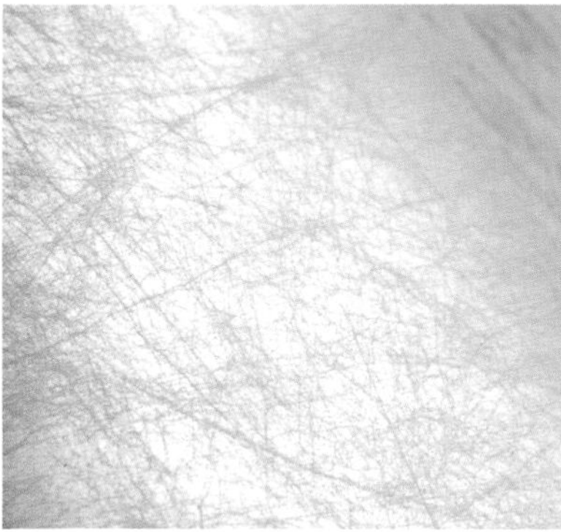

MC# 6799-04
Twist
Favini S.r.l.
www.favini.com

FSC-certified papers generated from renewable resources that are produced in accordance with environmental certification ISO 140-01. The paper has a textured multi-linear finish that can be easily cut, folded, and printed in twelve different colors. **Applications** typically include packaging and printing media.

MC# 6867-01
GGP Eco-Friendly Food Containers
Chang HWA Industrial Co., Ltd
www.ggpack.kr

A disposable packaging and serving-ware material composed of SiO_2 (sand) and polypropylene (PP) that is food-safe and microwavable. This composition creates a packaging material that uses fewer petrochemical-based plastics when compared to PP-based containers in general. **Applications** are primarily for the packaging and storing of food, but it can also be employed to package electronics.

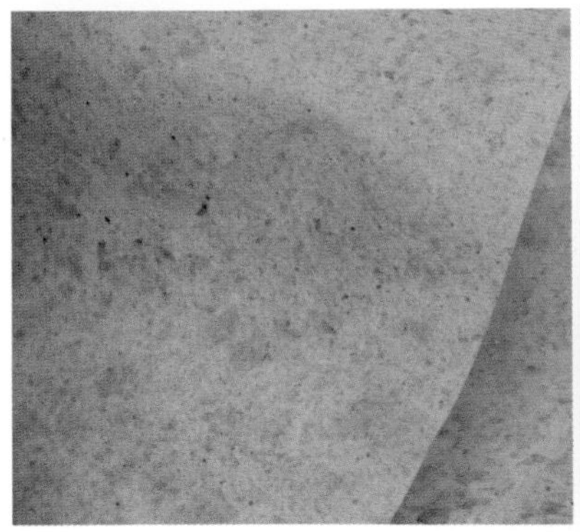

MC# 6918-02
Seaweed Paper
Sum Art., Ltd
www.tamnagi.co.kr

Handmade paper that contains natural seaweed pulp. The paper is fabricated with a mixture of natural pulps such as mulberry, wood pulp, and seaweed powder, blended with local natural components, in this case seaweed pulp. (Blends are 50–95% seaweed and 50–5% pulp.) The material can be processed in all the same manners as standard paper. **Applications** include packaging, stationery, consumer goods, accessories, wallpaper, and interior decor.

MC# 6922-01
BRELLIX
BRELLI
www.thebrelli.com

100% biodegradable PVC sheets that are phthalate-free, filter out 99% of UVA/UVB/UVC light radiation, and are printable. They are available in transparent or opaque and can be dyed any color. They also come with a double-polished clear texture or any custom-embossed texture per client request. The material meets ASTM standard 5511 for determining anaerobic biodegradation of plastic materials and CONEG (Coalition of North Eastern Governments) for regulating heavy metals formulated to filter out UVA/UVB/UVC rays. **Applications** include apparel and accessories.

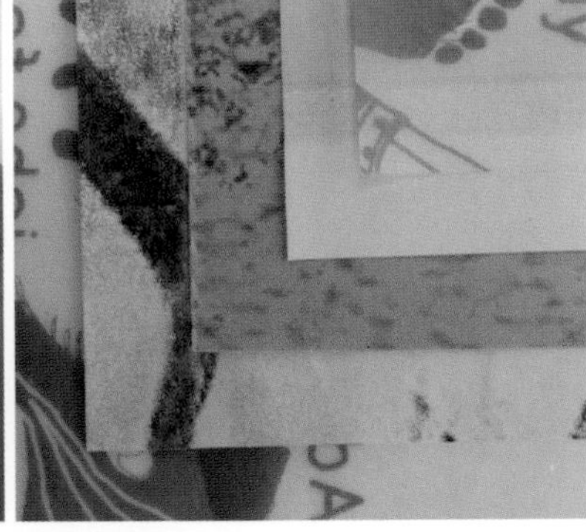

MC# 6988-01
Ecoplas
Ecopia Worldwide, LLC
www.goecopia.com

A flexible biopolymer packaging composed of 60% tapioca starch and 40% polyethylene (PE) that is biodegradable, washable, and durable. Tapioca is grown organically using native farming methods and is a non-genetically modified organism (non-GMO). The plastic can be processed using traditional blow- and injection-molding processes. The blown film is available in a variety of thicknesses. **Applications** include reusable totes, single- or multiple-use bags, transport bags, gift/promotional bags, table covers, packaging, hangers, hooks, and fixtures.

MC# 7000-02
Terratek® BD
Green Dot Holdings LLC
www.greendotpure.com

A compostable biopolymer comprised of natural and synthetic polymers, the base of which is starch from grain sources. This biopolymer exhibits improved heat tolerances over that of most biobased plastics, with a heat deflection temperature (HDT) of 92.8°C (199°F), as well as rigidity balanced with moderate flexibility. This polymer is compostable in accordance with industry standard ASTM D6400. **Applications** include packaging materials for products and food, disposable cutlery, home goods, toys, and consumer products.

MC# 7000-03
Terratek® SC
Green Dot Holdings LLC
www.greendotpure.com

A blend of wheat starch and polypropylene (PP), this resin minimizes polymer usage by loading the resin with starch. Biobased content offers the benefit of decreasing the amount of non-renewable resources used, and the polymer composite can be utilized in pliable and rigid applications and withstand temperatures above 100°C (212°F). Variable degrees of starch loading offer different physical properties of the blend, varying tensile strength and flexibility. **Applications** include food packaging, accessories, consumer goods, and household items.

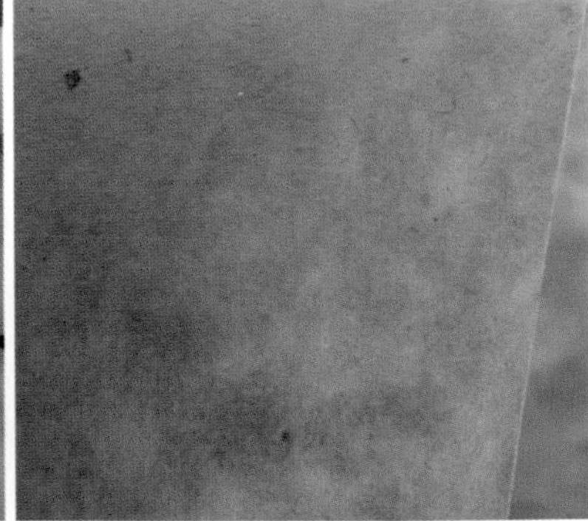

MC# 7054-01
Recycle-Compatible PURE Labels™
Distant Village
www.distantvillage.com

Labeling material composed of Fair Trade paper pulp from sustainably harvested hemp (40%) and wild grass (60%) backed with an adhesive. This recycling-compatible adhesive is an improvement over most commercially available label materials because it does not interfere with the paper-recycling stream. Wild grass and hemp fiber are harvested, shredded, and mixed into a pulp. The paper slurry is dried in the sun, and then the adhesive and liner are applied to convert it into a label. **Applications** include labels for gourmet and specialty foods, natural health and beauty products, retail packaging, and promotions.

MC# 6774-04
FOLEC-ST
INOS TECHNOLOGIES
www.inostech.com

Strong and rigid closed-cell microcellular polyolefin foam manufactured through a proprietary ultra-clean foaming process. The process uses neither chemical foaming nor crosslinking agents and has low outgassing and residual contamination values. The foam has cell sizes from greater than 100 microns to less than 200 microns. Colors are black and white. **Applications** include building and vapor control.

MC# 7036-01
Skalax™
Xylophane AB
www.xylophane.com

A xylan-based barrier coating for cardboard and paper products that protects against oxygen, grease, ink, and scent transmission. Biodegradable and made from rapidly renewable resources, this type of coating can add functionality and safety to biobased, recyclable, and compostable materials while preserving their low environmental impact. Xylan, a carbohydrate, is extracted from agricultural by-products, then mixed with proprietary additives, and finally layered onto a paper or board surface through dispersion coating. **Applications** include packaging for food and consumer goods.

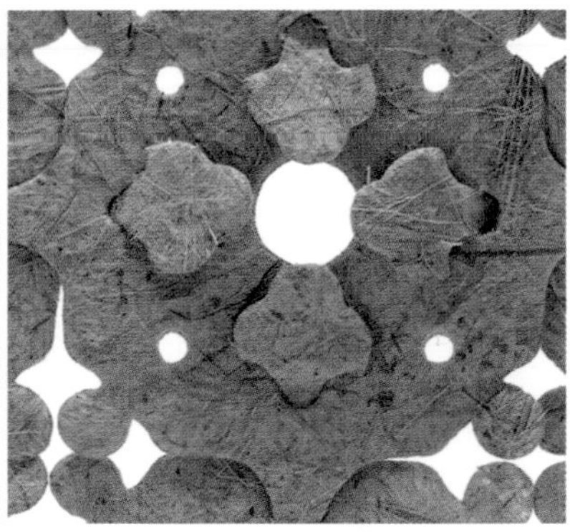

MC# 7100-01
Ma-Lee (Banana Paper)
SARRAN

Undyed paper composed of 98% banana fibers, 1% banana mucilage, and 1% carbon powder. Created using a proprietary process, this material does not use additional chemicals, is biodegradable, absorbs odors, heat, sound (as insulation), and perfume, and is water-resistant and durable (it has 30% higher tear resistance than regular paper). The clean banana fiber is mixed with carbon powder, banana mucilage is added, and the mix is molded into paper. **Applications** include room dividers, partitions, interior decoration, wall coverings, place mats, and rugs or carpets (floor decoration).

MC# 7115-01
Step Forward Paper
Prairie Pulp & Paper Inc.
www.stepforwardpaper.com

This is the first wheat-straw-based copy paper on the market, and uses waste straw that does not affect the food chain. It is a high-quality copy paper that uses predominantly non-wood waste fiber resources, combining 80% wheat straw and 20% FSC-certified hardwood pulp fiber. The straw is washed and cleaned, with pulping done in a continuous digester. It is bleached using elemental chlorine-free methods, mixed with the hardwood pulp, and transformed into paper. **Applications** span home, office, and commercial printing.

MC# 7200-01
Evlon® Compostable Plastic Film
Bi-Ax International Inc.
www.evlonfilm.com

Biaxial film made from NatureWorks™ biopolymer of polylactide (PLA) resin intended as a replacement for petroleum-based plastic films. This material is clear, made from renewable resources (corn), and is industrially compostable. It has comparable properties to current plastic films, including high stiffness, high gloss, high strength, and excellent twist retention and dead fold (it does not unfold spontaneously). It is an excellent flavor and aroma barrier, and is resistant to water, grease, and oils. **Applications** include vertical and horizontal filling and wrapping machines, and in packaging of twist wrap, window film for bags or boxes, board lamination, and pressure-sensitive labels.

MC# 7229-01
Three-Dimensional Hanji with Embossed Pattern
Shin Pung Hanji
www.museumhanji.com

Handmade Korean dak-mulberry paper with a three-dimensional pattern. Hanji paper (traditional Korean handmade paper) is known for its durability and strength; the paper absorbs odors and moisture, can be sterilized, and is insulating. Silk is added to the paper surface after coating with a grain powder; embossing the silk on to the Hanji, it reacts with heat and the grain inflates and makes the three-dimensional effects. **Applications** include interior decorations, craftwork, filters (masks/purifiers), and shading textiles.

MC# 0149-02
ECOMAGIC
Kuraray America Inc.
www.kuraray.co.jp

Enabling easy open and close, this hook-and-loop fastener is produced in a solvent-free process. In comparison to conventional hook-and-loop fastening systems, it does not have a polyurethane (PU) back coating and the homogeneous 100% polyester composition is easily recyclable. The material is colorfast and non-yellowing and can be used in vulcanized constructions, and the peel and sheer strength of the fastening system remain the same in wet conditions. This material can be stitched or adhered to textiles or solid surfaces. **Applications** include fasteners for footwear, apparel, bags, and accessories.

MC# 3086-05
Ahlstrom EasyLife®; Textile-like Lined Uncoated Facing BR5380
Ahlstrom
www.ahlstrom.com

Providing a pronounced ridged surface texture, this non-woven paper can be printed with rotary screen-printing processes. It has high dimensional stability and can easily be hung and removed from walls, which is its primary use. This material is composed of 65% wood fiber (FSC certified), 20% synthetic fibers, and 15% synthetic binders, and produced with a specialized wet-laid non-woven process. It is available in a standard white color and can be printed on the top (ridged) face with Hewlett Packard Latex technology and water-based screen-printing processes. **Applications** include paper stock for printed wallpaper and wall coverings.

MC# 5099-02
NanoWave
Holingsworth and Vose
www.hovo.com

Polypropylene (PP) non-woven textile with a pleated interlayer for improved filtration properties. The various fiber sizes and internal pleating increase the surface area of the textile as compared to a flat textile of equivalent thickness, offering a significant improvement on other synthetic-fiber filters, with double the dust-holding capability. The textile can then be heat- or ultrasonic-welded into final product shapes. **Applications** include filtration of gas media, respirators, residential and commercial HVAC systems, vacuum-cleaner bags, and exhaust filters.

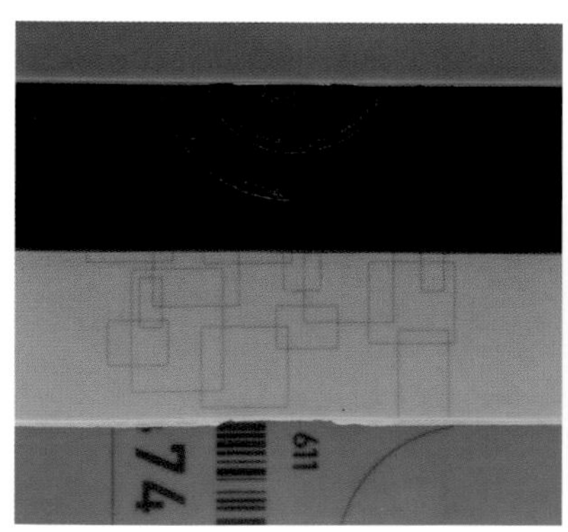

MC# 5243-03
Iriotec® 8000 Series
EMD/Merck
www.merck-pigmente.de

Enabling the clear, permanent laser-marking of plastics, these functional pigments enhance the marking speed, contrast, and resolution of all common plastics, including difficult-to-mark plastics such as polyethylene (PE), acrylonitrile butadiene styrene (ABS), polyamide (PA), and polycarbonate (PC). The pigments are suitable for both light and dark plastics, do not significantly affect the "inactivated" color of the plastic, and are capable of producing visible contrast markings at low laser intensities. **Applications** include food packaging, cable sheathing, electronics (keys), and ear tags for the farming industry.

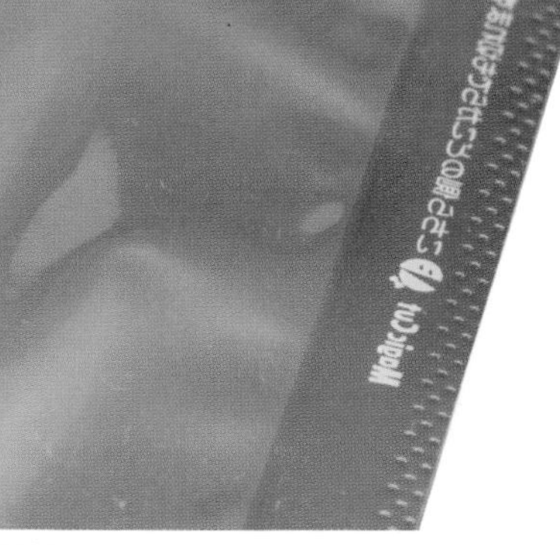

MC# 6142-07
Magic Cut
Asahi-Kasei Corporation
www.asahi-kasei.co.jp/asahi/en

A slitting or perforating technology for sealed seams of polymer films that make the seams easy to tear across. The continuous "scarring" of the film along the entire seam allows for easy opening at any point. A special blade is installed into the converting machinery that cuts, folds, and seams roll goods into finished products. This process can be applied to polyethylene (PE), polypropylene (PP), polyester (PET), polystyrene (PS), and nylon (PA) films, and to foils. **Applications** include packaging for food, products, medical devices, and toiletries.

MC# 6472-04
Miracure®; Silver
Henkel
www.henkel.com

A silver UV-curable alternative to foil-board laminating and hot foil stamping. It has a higher ratio of pigment in the formula and this gives the coating more of a matte finish. The coating can be applied to precise areas of the packaging. Since the ink is non-conductive, papers printed with this product will not be inadvertently removed from paper-recycling streams or block RFID tags used for security and asset tracking. **Applications** for packaging include folding cartons, labels, direct mail, and POP displays.

MC# 6455-02
PRESS-LOK®
Velcro® Brand Products
www.velcro.com

Flexible re-closable fastener composed of polyethylene (PE) self-mating micro-hooks. This closure does not require precise alignment in order to fasten; there is a tangible and audible cue that it has sealed; and unlike other closures it is light yet rigid so the package top does not fold over and any graphics remain flat and legible. The tape can be heat-sealed to polymer films and is capable of running on automated packaging machines. **Applications** include closures for flexible packages for dry or powdered food, industrial, medical and consumer products.

MC# 6816-01
TriPly
Sohner Plastics
www.sohner.com

A lightweight, three-layer, 100% polypropylene (PP) panel consisting of a central formed structural sheet sandwiched between two smooth exterior sheets. The wall thicknesses of the exterior sheets have good consistency, adding strength to the welded bond of the structural middle sheets and improving rigidity. Three finishes are available depending on the desired application: soft/smooth, non-slip, and textured/colored. The manufacturer can use virgin or recycled polypropylene to produce the panels and they are 100% recyclable. **Applications** are for packaging.

MC# 6830-01
Visualize
HLP Klearfold
www.hlpklearfold.com

A process for the accurate and repeatable folding of clear polymer cartons using radio-frequency (RF) heating. The process can be conducted on a wide variety of materials and achieve a vast range of styles, structural shapes, and sizes. It can also be used to create decorative and customizable packaging for specific products as well as branding This proprietary scoring technique uses radio frequencies to heat and bend the film in this packaging process to achieve cleanly folded and shaped durable packaging. **Applications** are for packaging.

MC# 6925-01
ULTRASTOP Foam
Ultralon Products (NZ) Ltd
www.ultralon.co.nz

Closed-cell foams that dissipate impact energy and prevent bouncing. They are comprised of chemically cross-linked polyolefins, either polyethylene (PE), ethyl-vinyl-acetate (EVA), or a blended copolymer. The foams are constructed in a molded bun form as a large slab, which can be cut to required size and thickness. The foams can be processed with standard foam-working equipment, and some wood-working tools. **Applications** include marine padding, automotive vibration pads, impact sound insulation, orthotic supports, protective sportswear, footwear, gymnasium mats, backpack padding, and toys.

MC# 6978-01
Polyfloss
The Polyfloss Factory
www.thepolyflossfactory.com

100% recycled polypropylene (PP) that has been processed to create a fiber that is easily remolded. This unique processing method is done to the recycled plastic and is a solution to the existing problem of separating composite materials before recycling. The plastic is first shredded then inserted into a rotating oven. Through centrifugal force, the molten plastic is projected through small holes on to a drum. The space between the oven and the drum lining allows the polypropylene to cool down and harden, creating fibers. The material can be easily remelted to create new objects that share different qualities. **Applications** are for fashion, packaging, and products.

MC# 7103-01
Multiflex Tubes®
Linhardt GmbH & Co. KG
www.linhardt.com

Process for creating seam-free tubular packaging by welding the edges of the flat material in a butt joint and sealing with a thin tape, rather than conventional overlapping and welding techniques. This process allows for 360° printing of the entire tubular surface, partial spot lacquering, and the asymmetrical combination of layers within the composite packaging material. It is composed of decorative, barrier, and functional layers, is easily formable, and has very good oxygen- and moisture-barrier properties. **Applications** include primary packaging for creams, gels, and ointments for personal care, oral care, pharmaceutical, food, and household and industrial products.

MC# 2984-04
Easy-Lock
APLIX Inc.
www.aplix.com

Flexible transparent fastening tape containing rows of self-mating micro-hooks. Composed of food-grade low-density polyethylene (LDPE), this fastener is re-closable and resistant to contaminants and does not require precise alignment in order to fasten. The tape can be heat-sealed and is capable of running on automated packaging machines including horizontal and vertical form, fill and seal (FFS), and pouch-making machines. The polyethylene used in this material has been approved for contact with food by the US FDA, the CFIA, and the European Commission. **Applications** include closures for the packaging of food, industrial, and consumer products.

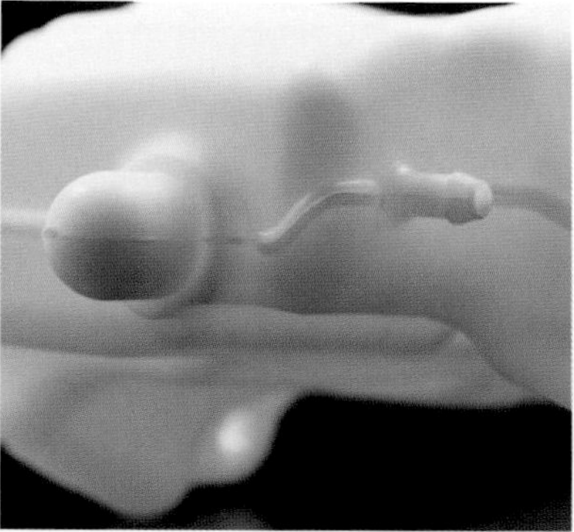

MC# 5381-04
GAS-SOSY©
JSP
www.jsp.com

Plastic blend that is suitable for the blow-molding of fuel tanks. This combination of high-density polyethylene (HDPE) and polyamide (PA; nylon) creates a mixture that has good gas- and fluid-barrier properties with no additional film barriers or fluorination treatments. It passes EPA 40 CFR Part 1060.520 for the control of evaporative emissions (the tank does not leach gas emissions). It is suitable for fuel tanks as well as other containment tanks that require high-performance barrier properties.

MC# 5479-02
Molded Fiber
UFP Technologies
www.ufpt.com

100% post-consumer recycled flexible molded pulp-based packaging material system. It can be used to create a variety of different custom-molded and engineered shapes including clamshells, trays, inserts, end caps, and other protective products. These engineered shapes can be made to place emphasis on cushioning and protecting a product, as well as being nestable and stackable. Each run is customized for the particular project, and customized colorization and finishes are possible. **Applications** range from health and beauty to cosmetics packaging.

MC# 6087-02
Edible Films
MonoSol, LLC
www.monosol.com

Clear and colored biodegradable polymer films that may be safely eaten. These polyvinyl alcohol (PVOH) -based resins are FDA approved and are designed to hold the contents of packages of food and maintain a good moisture barrier until immersed in water. They dissolve completely in cold water with no residue and are safe for handling by consumers. This type of packaging is ideal for flavors, colorants/dyes, enzymes, vitamin fortifiers, conditioners, yeasts, and drink mixes. **Applications** include single-use food and beverage additives.

MC# 7133-01
Versalite™
Berry Plastics Corporation
www.versalite.com

Unlike other hot beverage cups that can use multiple materials, this version is completely recyclable. Foamed polypropylene (PP) sheets are converted into durable insulating drinking vessels suitable for hot and cold beverages. Each sheet is foamed, with a smooth outer and rough inner surface; the outer surface can be easily printed with graphics using 10-color HD flexographic printing, and the sheets are typically 1.75 mm (0.07 in.) thick and in continuous lengths. A sheet can be produced in any color, but comes in a standard white.

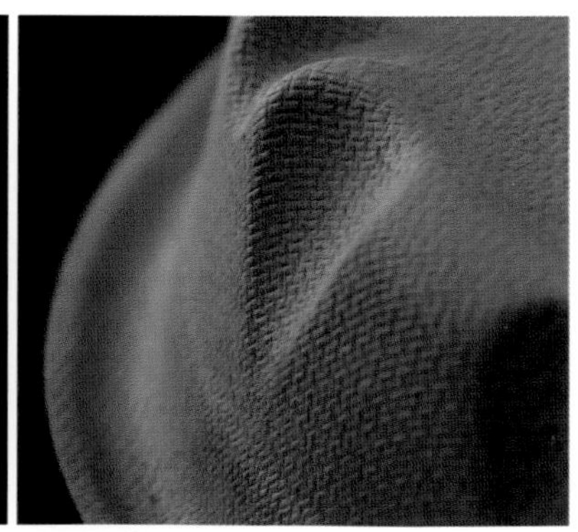

MC# 7255-01
Ultra Green™ TreeSaver™
Ultra Green Packaging Inc.
www.ultragreenhome.com

Moldable non-wood paper pulp composed of renewable natural fibers (wheat straw or bamboo) and FDA-approved additives. This material is intended as an alternative to plastic, foam, and aluminum for disposable applications. It is durable, oil- and moisture-resistant, compostable, and biodegradable; it can be microwaved and baked at temperatures up to 204°C (400°F). **Applications** include food containers, packaging, and consumer products.

MC# 6921-01
Airolux
Airopack
www.airopack.com

A patented pressure-control technology that enables the dispensing of fluids and other medium- to high-viscosity fluids. This system replaces traditional aerosol products used with conventional metal cans and chemical propellants. It is able to achieve a standard 2.0-bar pressure (for creams and gels) as well as the high pressure (3.5 bar) necessary for sprays. The entire system is manufactured from polymer resins with no need for metal parts, with immediate dispensing. It uses compressed air rather than hydrocarbon chemical propellants, enabling the pressurized dispensing of products not previously possible.

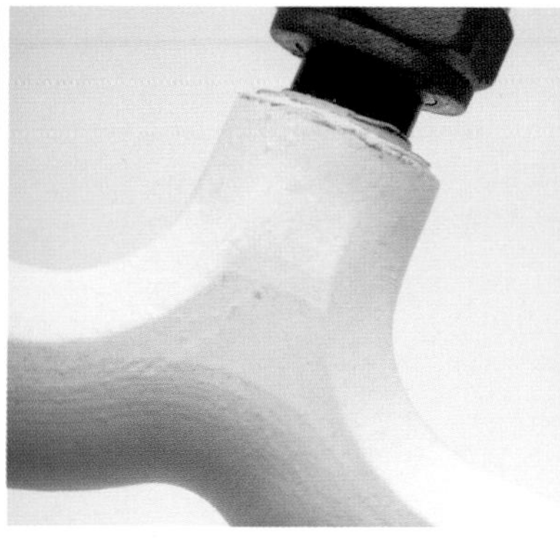

MC# 6950-01
ecooler system
StudioKahn
www.studiokahn.com

Ceramic cooling system that does not require electricity. The system cools down the air by seepage and evaporation of water through the clay. Water-filling can either be performed automatically or manually; a visible indication of the decrease in water level will notify that the system needs to be filled (the color of the clay and its feel will vary when wet or dry). The ceramic pieces are non-flammable and resistant to all weather conditions. It is available in a natural white color. **Applications** include climate and temperature control.

MC# 7039-01
Shape Retaining Plastic
Sekisui Seikei, LTD
www.inabataamerica.com

Plastic sheet that has the formability of soft metals and wire, and can be shaped repeatedly by hand in multiple directions. This allows for single-material products with increased ease of recycling, providing the same functionalities as multi-material products. Multiple lamination options for customized directional pliability can be made. Additionally the surface is suitable for printing. **Applications** include food packaging, packaging, fashion, accessories, custom fit bands, sports protection, and nosebands.

MC# 7158-01
Magnetic Strip
Paskal PK Zippers & Fasteners
www.gogooper.com

Waterproof, self-sealing magnetic closure composed of strong neodymium magnets and a durable, flexible thermoplastic polyurethane (TPU) strip. When the strips are connected to one another they create an instant, waterproof seal, and the TPU fully encapsulates the magnets, so they will not rust. This eliminates the need for a snap-and-lock or a hook-and-loop system to create an airtight closure. **Applications** include fasteners and waterproof bag closures.

MC# 7159-01
No. 2 Pouch
Ampac
www.ampaconline.com

Packaging made from coextruded high-density polyethylene (HDPE) and linear low-density polyethylene (LLDPE) that is suitable for recycling in municipal programs. Recycling streams are often not able to accommodate polymers in thin-film and bag forms; this pouch is approved for recycling as HDPE. The film provides puncture resistance, a moisture barrier, and grease resistance. HDPE and LLDPE pellets are melted down and extruded into a three-layer construction, with one layer of LLDPE on either side of an HDPE center layer. The films are cut and heat-sealed into pouches. **Applications** include packaging for food and wet and dry goods.

MC# 7189-01
AquaFlexCan
Amcor Flexibles Europe and Americas
www.amcor.com

Flexible laminate film that can be converted into a stand-up flexible beverage container. The barrier properties of the film ensure that its contents, particularly water, remains fresh and pure; and its use as a container reduces waste in comparison to conventional plastic bottles. Features of the packaging include a laser-perforated opening from which to sip, requiring no straws or scissors. As a beverage container it is lightweight and collapsible once empty. **Applications** include packaging for non-carbonated beverages such as water, energy and sports drinks, juice, and soup.

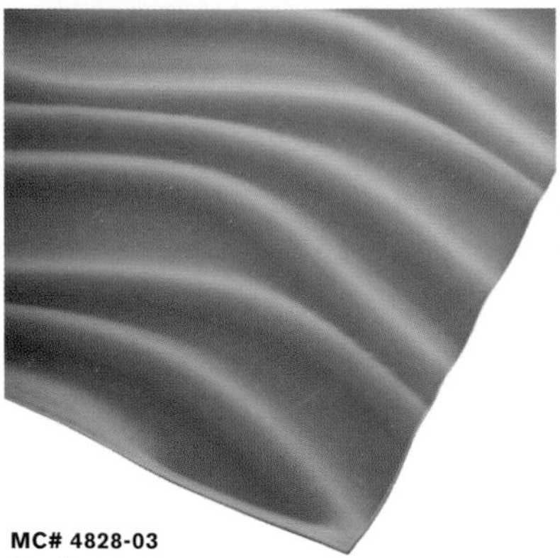

MC# 4828-03
EccoFlex
Advanced Technology Inc.
www.advtechnology.com

Laminate made from a thermoplastic alloy formulated to substitute for fire-retardant acrylonitrile butadiene styrene (ABS). The laminate is made of a polyvinyl chloride (PVC) and acrylic blend including 40–50% pre-consumer recycled content. The sheets are available in beige and black and are compliant with California Green Building Standards Code, LEED (for recycled content), and Green Globes. This material is lightweight and flexible and has high impact resistance. **Applications** include walls, ceilings, furniture, exhibits, displays, and counters.

MC# 5441-02
Microtaggant® Security Yarns
Microtrace LLC
www.microtracesolutions.com

Security yarns that incorporate uniquely encoded microscopic micro-tags for identification and security purposes. This yarn provides individual security features including ones specific to the time of product manufacture, intended market, manufacturer identification, authenticity verification, or yes/no verification. Identifiers unique to the product or manufacturer are traceable, and provide increased security detail over current systems that are uniform across products. **Applications** include anti-counterfeiting, tracking, and product protection.

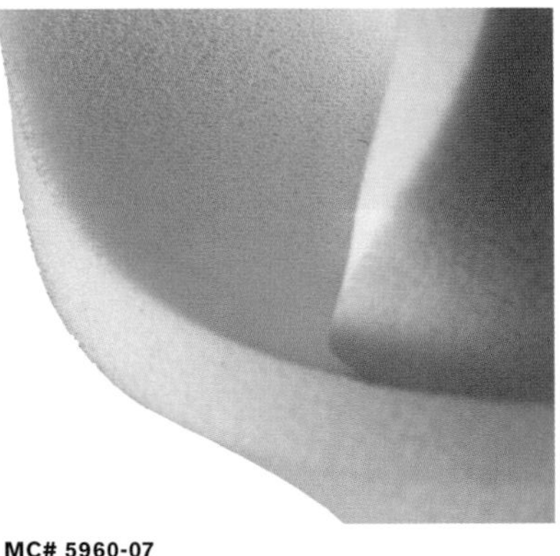

MC# 5960-07
SuperSoft HL
Filtrona Porous Technologies
www.filtronaporoustechnologies.com

Hydrophilic high-density polyurethane (PU) foam that has a super-soft surface that is capable of self-sealing. It is made from 50% PU and 50% other proprietary ingredients. It is processed through continuous sheet foaming and meets standards that are suitable for skin contact. The foam may be die-cut to any shape. **Applications** include cosmetic foams, cushioning, filter components, and other consumer products.

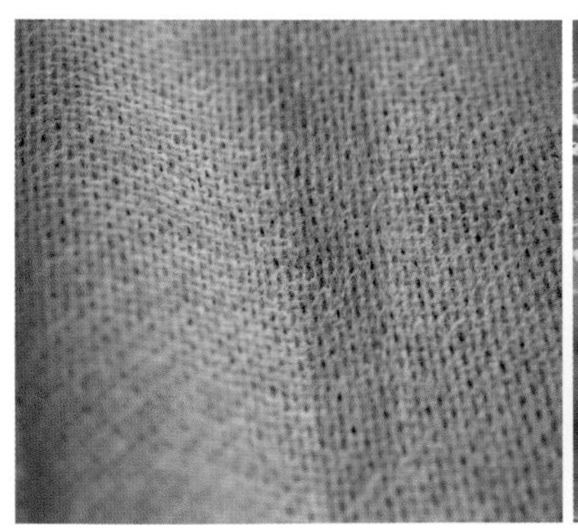

MC# 6142-06
Bemliese
Asahi-Kasei Corporation
www.asahi-kasei.co.jp/asahi/en

Continuous-filament cellulosic non-woven. Most non-wovens are composed of either natural staple fibers, which can lint or catch on objects, or synthetic fibers, which falter in liquid absorption, liquid retention, resistance to electrostatic charge, heat resistance, and biodegradability. This non-woven is unique in that it is composed of natural fiber filaments. Made with CuPro, a biodegradable cellulosic fiber with soft and silky hand, it has good liquid absorbance, thermal stability, pick-up, purity, and biodegradability. **Applications** include medical gauzes, clean room wipes, biodegradable agricultural and horticultural mats and nets, teabag filters, cosmetic cloths, and disposable wipes in hospitals.

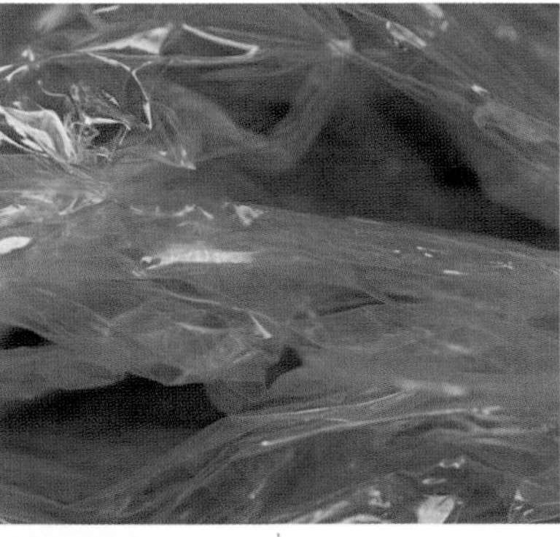

MC# 7209-04
Styroflex 2G-66
Styrolution America LLC
www.styrolution.com

A low-cost, high-performing styrene polymer that exhibits the properties of an elastomer (rubber-like), with high transparency and soft touch. The polymer has stretch and strength, having an elongation at break of 550%. Blow-molding and cast extrusion are used with this polymer for film formation as well as extrusion for the production of sheets and profiles. The polymer is certified for use in medical devices and is suitable for use in articles indented for food contact. **Applications** include multilayer films, food and product packaging, industrial packaging, and modification of other polymers through compounding.

MC# 7267-02
Flame Retardant deTerra® Biobased Polymer (PR 146)
Interfacial Solutions (IFS) Engineered Plastic Compounds
www.interfacialsolutions.com

Extrusion-grade resin made from 93% polylactic acid (PLA) biopolymer, derived from renewable resources such as corn, and proprietary non-halogenated flame retardants. The material is compostable, can achieve a V0 or V2 fire rating (under the UL 94 flammability standard), and has improved properties over conventional PLA, including excellent melt strength and good ductility. The PLA content of this material can be sourced from post-industrial scrap and is intended as a "green" alternative to polyvinyl chloride (PVC). **Applications** include building and construction materials, and consumer goods such as ink cartridges, electronics, and appliance casings.

MC# 7028-01
UFLEX
LM Packaging
www.lm-packaging.com

Custom performance packaging made from 100% recyclable paper that has been machined with a proprietary process to bend around corners at any point in the package. It has superior energy absorption and impact protection, and is lightweight and recyclable. This packaging is designed to specific customer requirements. It can be die-cut, scored, and laminated with bubble wrap on one or both sides for extra protection, and coated with anti-abrasion coating. **Applications** are for protective packaging for curved parts in the automotive, electronics, and medical industries.

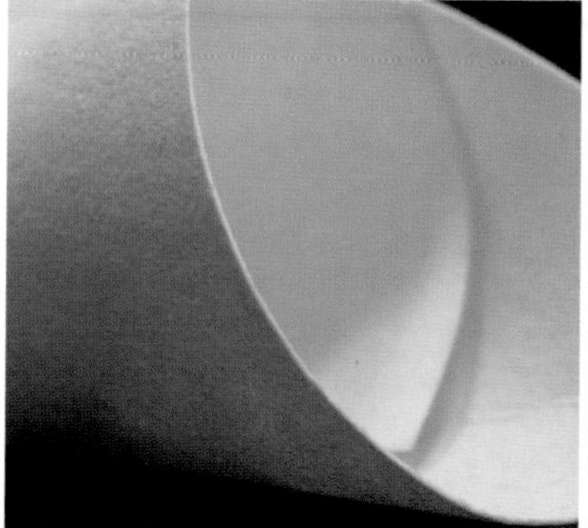

MC# 7061-01
Ceramicpaper Typ 3000
Kager Industrieprodukte GmbH
www.kager-hightemp.com

Inorganic, textured ceramic-fiber sheets made from a combination of raw materials used as insulation. The material is fire-resistant and heat-insulating, and the technology makes it possible to produce this fiber by the fusion of alumina and silica at high temperature. The paper is offered in off-white only and there is no customization available for this material, but it is available in either ceramic paper (including a bio-soluble variant), ceramic fabric (also including a bio-soluble variant,) ceramic fiberboards, or moldable ceramics as cement. **Applications** for this material include high-temperature insulation.

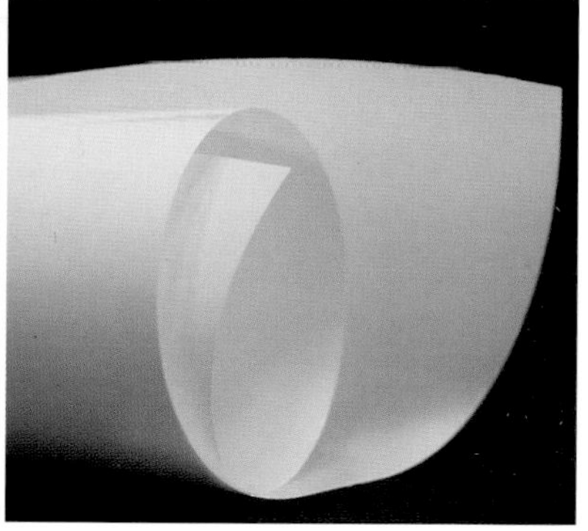

MC# 7074-01
Nor®Cell
Mondi Gronau GmbH
www.mondigroup.com

Thin, flexible, extruded-polyolefin film with a foamed middle layer. Compared to standard un-foamed polyethylene films, this material offers material and weight savings of 40% and improved insulation without changing the physical properties such as haptics, thickness, and stiffness. Composed of 90% polyolefin and 10% filler, a patented extrusion-blown film process produces micro-cells within the film's core layer. **Applications** include flexible packaging, lamination structures, pre-made plastic bags, surface-protection films, and printed or non-printed films for the food, pet food, and hygiene industries.

MC# 7075-02
Ultra Ever Dry
UltraTech International Inc.
www.spillcontainment.com

A two-part coating that creates water- and oil-repellent surfaces on a variety of substrates, such as metals, concrete, wood, leather, and textiles. Unlike other coatings for repellency, this features simple application via spray, using air sprayers, pump sprayers, and even finger-trigger sprayers. Air-drying is sufficient, though a heat dryer will speed the process. The contact angle of the surfaces, post-treatment, is 165–175°. It is fluorochemical free, using silica as the surface modifier. **Applications** include parts for motor vehicles and machinery, building exteriors, boats, footwear, clothing, and electronics.

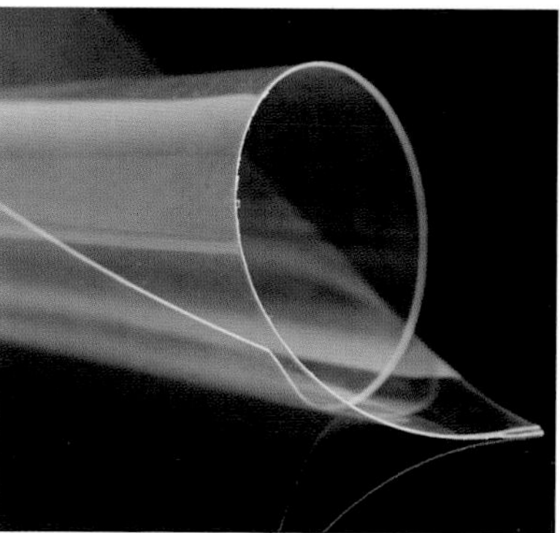

MC# 7126-01
Meliflex
MELITEK A/S
www.melitek.com

An alternative to vinyl for blood bags, this high-clarity three-layer polypropylene (PP) flexible film is free of latex, BPA, plasticizers, and phthalates. This film offers high-performance physical and optical properties that are usually only found in plasticized vinyl (PVC). It provides good chemical resistance, impact resistance at low temperatures, low sealing temperatures, and can be sterilized in an autoclave. Three grades of PP resin are simultaneously extruded through a three-layer die head at different thicknesses to form a sealing layer, a core layer, and an exterior layer. **Applications** include blood bags, medical devices, packaging, healthcare accessories, and tubing.

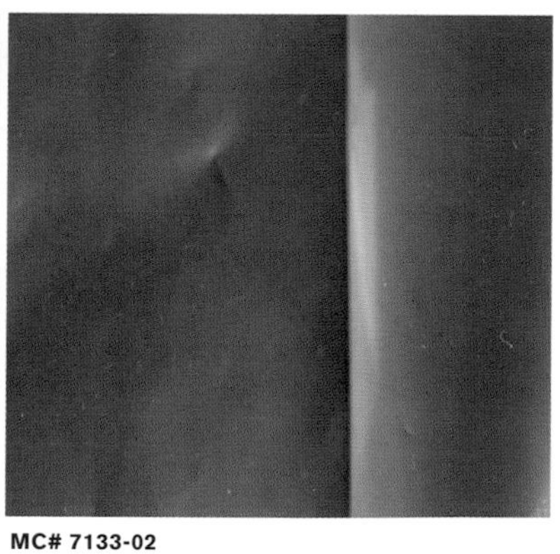

MC# 7133-02
Novel 9-Layer Film
Berry Plastics Corporation
www.versalite.com

Transparent, flexible film with nine discrete layers giving enhanced properties and functionality. Traditionally, films for food packaging have been limited to five or seven layers, with different polymers providing the desired functionality. This technology has been improved because it includes nine layers and highly engineered polymers to precisely manage all aspects of the finished film's properties and performance. Fourteen different resins are combined into the available nine layers in the coextruded film maximizing the performance of each raw material. **Applications** include packaging for dry foods, but the formulations could be customized for a variety of foods.

MC# 7152-01
Skydex
SKYDEX Technologies Inc.
www.skydex.com

Cushioning system that uses an array of collapsible square cup structures to give support. This system, which can use different-sized cups to create different thicknesses and comfort levels, offers an improvement over other cushioning materials such as foam as it can be easily washed, has greater durability (up to 10 times as long), and can also be recycled. It has been tested for military use and for impact absorption by the American NFL (National Football League). **Applications** include football helmet cushioning for mattresses, impact absorption for flooring in aviation, and as blast mitigation in rapidly deployable structures.

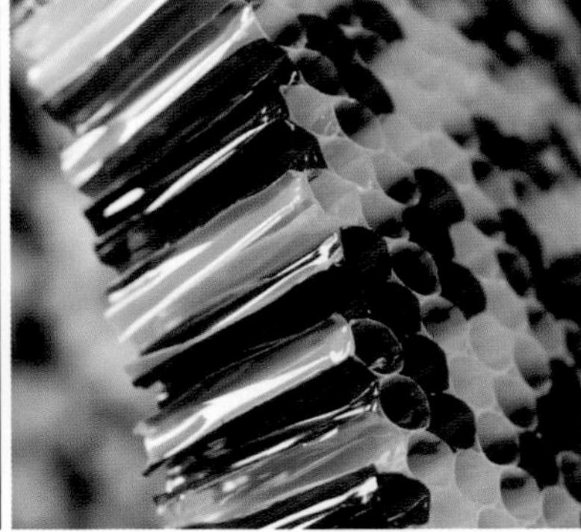

MC# 7164-01
Koroyd®
Koroyd SARL
www.koroyd.com

Lightweight shaped impact absorber. An engineered thermoplastic honeycomb comprised of a coextruded polycarbonate (PC) provides uniform mechanical properties thanks to its circular cell structure, and offers high compressive strength in a low-density material, decreasing transmitted force and peak g-force acceleration. The honeycomb is an efficient energy absorber, which is vital to impact protection, and is highly breathable. **Applications** include helmets, sports equipment, body armor, blast mitigation, automotive panels, board-sport boards, furniture, lighting, displays, and interior decor.

MC# 7208-02
Tylac® 4193
Mallard Creek Polymers
www.mcpolymers.com

Latex emulsion comprised of styrene-butadiene copolymer with high flexibility and softness. This fluid imparts elasticity to systems it is added to, and retains plasticity in temperatures as low as −15°C (5°F). The increase in plasticity results in increased strength, as pliant structures have higher resistance to brittle failure. The fluid also creates water/ion-resistance, and when added in large amounts the emulsion creates a damp-proof membrane. **Applications** include concrete modification and flexible membranes.

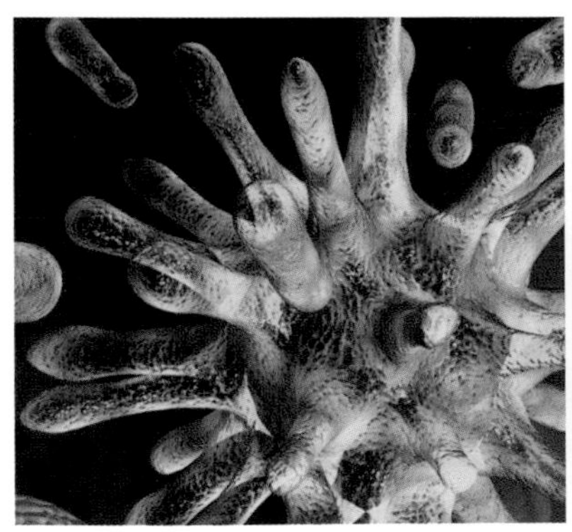

MC# 7241-02
Bacterstop®
Derprosa Film
www.derprosa.com

Antibacterial orientated polypropylene (OPP) film. The composition of the film kills 99.9% of bacteria that comes into contact with its surface. The lamination provides a barrier against any contagion. The film also has excellent bonding, UV varnish (matte or glossy), and stamping capabilities. The film is transparent, and has no color. Customization is available for sizes. **Applications** include brochures, leaflets, packaging, and books.

MC# 7241-03
GSP (Gluable Gloss)
Derprosa Film
www.derprosa.com

Clear, glossy film, composed of biaxially-oriented polypropylene (BOPP), which is gluable, stampable and printable. Intended as an alternative to acetate, high gloss biaxially-orientated polyester (BOPET), and acrylic coated films, it has excellent machinability and bonding properties to paper, board, and film, and is 20% brighter than standard gloss films. It is suitable for cold adhesives and can also be embossed or printed with a raised texture; glitter can also be adhered to the surface. **Applications** include magazine and book covers, catalogues or brochures, and packaging (bags and boxes) for luxury goods.

MC# 7262-01
Constellation Jade
Fedrigoni S.p.A.
www.fedrigoni.com

Paper and paperboard that has exceptionally high abrasion resistance. It is composed of long cellulosic fibers from FSC-certified sources. One side of the paper is coated with a pearl-effect pigment and then off-machine embossed for decorative effect; the papers are resistant to acetone and abrasion, and are suitable for low-temperature use. The paper and boards can be cut, die-cut, scored, folded, or glued. **Applications** include greeting cards, labels, box liners, book covers, catalogues or brochures, and packaging (bags and boxes) for luxury goods, and "skinning" goods made from plastics or metal.

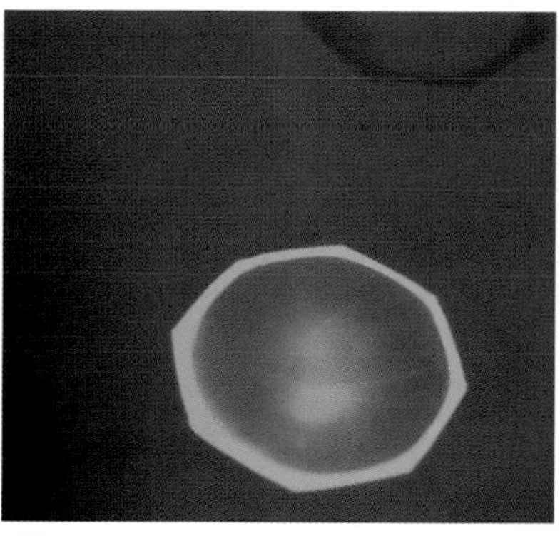

MC# 2604-26
Popper
Sommers Plastic Products Co. Inc.
www.sommers.com

Unlike standard color-change materials, this version offers two distinct color changes, whether the room temperature is increasing or decreasing. The fabric is a durable 98% polyurethane (PU), 2% proprietary liquid-crystal formulation (cholesterol-based organic coating) with a polyester textile backing. In this fabric, at 18–30°C (64–86°F) typical room temperature, the color is tan; below 18°C (64°F) the color is a dark olive; and above 30°C (86°F) the base color "pops" out, in this case a bright lime green. **Applications** are for footwear, apparel, accessories, toys, business cards, and promotional items.

MC# 2771-05
YUPO Octopus
Yupo Corporation
www.yupousa.com

With its ability to stick impermanently to almost any smooth surface, this synthetic printable paper uses a technology called "micro-suction." Extruded from 100% polypropylene (PP), it is completely waterproof, tear-resistant and stain-resistant, and does not require surface treatment before printing. It has been awarded the highest label (three stars) for substrate certification by Hewlett-Packard. Glass, PVC, aluminum, coated paper, and other smooth surfaces will work well with this synthetic paper. Since it does not rely on magnets or static, it is safe for use on electrical surfaces without adverse effect. **Applications** are for labeling and packaging.

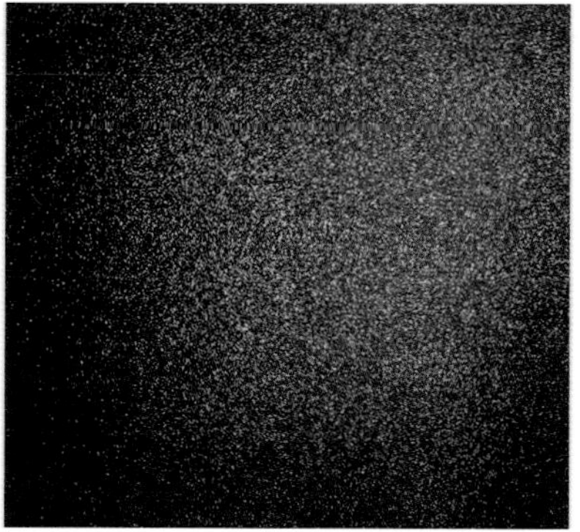

MC# 2604-27
Spectral
Sommers Plastic Products Co. Inc.
www.sommers.com

Ultra-reflective, brightly colored two-sided iridescent sheet composed of three laminated layers of film: 100% thermoplastic polyurethane (TPU), 3M Radiant Film, and Cube Light Reflective thermoplastic polyurethane (TPU). The textile is translucent but also has the ability to act as a reflective medium for color depending on the type and angle of the light source. Red, orange, magenta, and blue are some of the color frequencies given. **Applications** include displays, footwear, costumes, and accessories.

MC# 2771-06
YUPO® Reveal
Yupo Corporation
www.yupousa.com

This tamper-evident label made from synthetic paper is composed of polypropylene (PP) resins, inorganic filler, and additives. When the label is peeled, a customized non-sticky printed "VOID" message will appear on the surface of the product. The label cannot be reapplied to the product after removal and the layer remaining on the product is difficult to remove. This material gives flexibility to converters (companies producing packaging from roll and sheet stock) to use different designs instead of the traditional "void" image. A biaxial stretching method creates innumerable micro-voids that give opacity and whiteness by diffusion light reflection. **Applications** include anti-counterfeit, anti-shoplifting, tampering, and coupon labels.

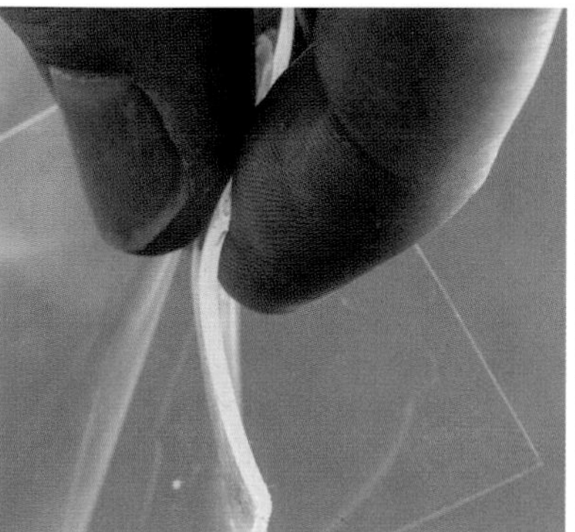

MC# 5086-17
OPTαGEL
Taica Corporation (formerly Geltec Co., Ltd.)
www.taica.co.jp/gel-english/

Typically used for touch panels, this optically transparent, soft silicone gel sheet with one adhesive face offers 99% light transmittance, making it exceptionally clear, comparable to glass. The material displays high shock absorption, while its thin profile and optical transparency allow for cushioning to be integrated within the screen. The gel sheet also displays durability against damage and yellowing from heat and UV exposure. The gel material improves luminance by 9%. **Applications** include shock insulation, optical parts, and luminance brighteners, for use in LCD and touch panel screens.

MC# 6973-06
Two-Shot Injection-Molding Decoration (IMD) of Plastics
Nissha USA
www.nissha.co.jp

Printed graphic film for two-shot injection-molding decoration (IMD) of plastics. This process imparts a sense of depth and three-dimensionality by layering graphics, a clear and a colored resin within the molded part. The IMD process replaces traditional decoration techniques such as painting, silk-screen printing, and pad printing. The surface of the part is protected by a hard scratch-resistant coating so that the films exhibit high durability (chemical, abrasion, and scratch resistance). **Applications** include casings for consumer electronics and home appliances, automotive interiors and exteriors, cosmetic packaging, and toys.

MC# 4051-02
Prescale®
Sensor Products Inc.
www.sensorprod.com

These thin films are used to measure the distribution and magnitude of pressure between any two hard surfaces. The low-cost, Mylar-based (PET) film contains a layer of tiny microcapsules that change in color intensity depending on the amount of pressure applied to the film. Applying force causes the microcapsules to rupture, producing an instantaneous and permanent "topographical" image of pressure variation across the contact area. The greater the pressure that is placed on the film, the more intense the color will be. **Applications** range across industrial and factory uses.

MC# 4194-07
PERMALIGHT® Power 250/30
American PERMALIGHT Inc.
www.americanpermalight.com

Glow-in-the-dark, photoluminescent vinyl sheets containing strontium oxide aluminate pigments that are non-toxic and non-radioactive and have the ability to be heat-formed. They are rigid, white to soft yellowish to yellow (depending on the pigment loading) in daylight, and glow greenish-yellowish in the dark. They are constructed of two layers: a top layer of luminous film and a base layer of white polyvinyl chloride (PVC). **Applications** include indoor safety signage to mark escape routes and obstacles, set design, and hospitality design.

MC# 4422-03
Reveal Inks
CTI
www.ctiinks.com

Enabling graphic and text reveals when the temperature changes, these thermochromic inks of the same color are printed adjacent to each other; they have the same activation temperature, but with differing deactivation temperatures, allowing a hidden message to be revealed. The color spectrum and intensity of these thermochromic dyes are superior to competitive inks on the market, and the color effect has the potential to enable brands to interact with their consumers post-purchase. Thermochromic inks contain microcapsules of dye and a developer within the same sphere. **Applications** include safety, security, novelty or promotional messaging for the packaging of food and beverages, labeling, toys, and wearables.

MC# 4422-04
Glow-in-the-Dark Inks
CTI
www.ctiinks.com

Printable ink, which is "charged" by outdoor or indoor lighting (whether natural or artificial), and then glows fluorescent greenish in dark environments or under black lighting. The ink has good rub resistance and a strong glowing effect, and the offset press formulation can be used straight out of the can as a spot color, rather than needing to go to an offline coater, lowering production costs. The ink is cream in color and is recommended for printing on a white or light background. **Applications** include safety, security, novelty or promotional messaging for the packaging of food and beverages, labeling, magazine inserts, posters, toys, and apparel.

MC# 4453-02
El Panel Power Shapes
M2 Lighting Solutions, LLC
www.thatscoolwire.com

Thin, flexible, electroluminescent (EL) panels that emit continuous ambient light across the surface of the panel with no "hot spots," opening up new options for seamless integration of lighting. EL material is sandwiched between two thin polymer film layers; wires are laminated at an edge for power. Two colors are available in all shapes, Cool White and Blue Green. Panels can be cut and customized to shape. **Applications** include wearable technology, signage, interior lighting, automotive lighting, transportation lighting, safety lighting, fashion, accessories, displays, and consumer goods.

MC# 5000-05
LIGHT LINE®
LEONHARD KURZ Stiftung & Co. KG.
www.kurz.de

This is a distinct hot stamping process made up of diffractive, continuous stamping foils, differentiated from other conventional techniques by the absence of shim lines in the final pressed foil product. Shim lines are fine seams that are created during the stamping process that detract from the homogeneity of designs. **Applications** for this graphic label stamping process include a wide variety of plastics, especially PVC cards.

MC# 5053-05
Wink
Wolf Gordon
www.wolf-gordon.com

Water-based finish that turns any paintable surface into a "dry-erase" surface. It performs as a writable finish like a whiteboard on any new or existing painted surface. Standard dry-erase markers are recommended. Writing can easily be removed with quality dry-erase cleaning sprays or wipes. **Applications** are for interior painted surfaces.

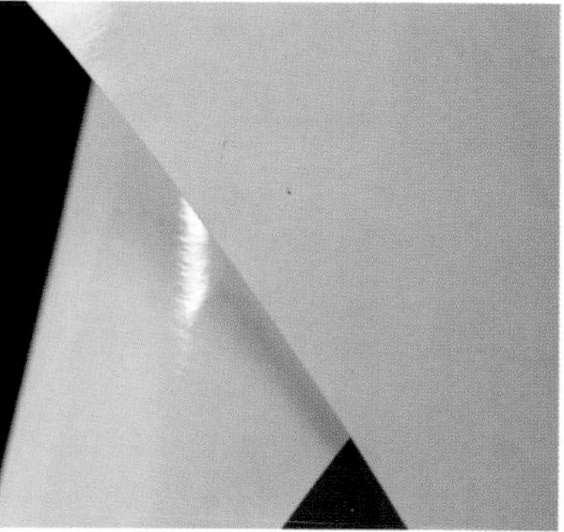

MC# 5429-16
PowerCoat®
Arjowiggins Creative Papers
www.curiousstory.com

Offering a chance for smart electronic packaging that is recyclable, this paper substrate is intended for high-performance printing of electronic conductive inks. It is 100% cellulosic paper and contains no plastics, but offers a polymer-like surface smoothness to enable improved ink conductivity, higher resolution, and reduced costs. The papers are biodegradable and recyclable in standard paper-recycling facilities. **Applications** include disposable labeling and packaging, RFID tags, resistors, capacitors, self-inductance and other passive components, lighting and display circuitry, as well as battery electrodes and sensors.

MC# 6992-01
Picasus®
Toray Industries Inc.
www.toray.us

Flexible polyethylene terephthalate (PET) film for in-mold decoration (IMD) with a metallic luster and appearance. Unlike films that contain metal, this film is recyclable, does not block electromagnetic waves, is free of cracks and blanching when formed, and has no possibility of corrosion. It offers environmental benefits as it can be used in place of traditional plating, coating, or metal vapor-deposition techniques. It is produced by a proprietary nano-layering technology that reproduces the sheen of metal without using any metal. **Applications** include casings for consumer electronics and appliances (mobile phones, music players, computers) and automotive trims.

MC# 7009-01
MicroMotion™
Technografix Inc.
www.Technografix3D.com

A patented clear acrylic coating that reflects light to create the illusion of motion. It is used primarily for visual effect, but provides abrasion resistance to most surfaces to which it is applied. The coating may be applied to almost any substrate (paper, plastic, glass, metal) by traditional printing techniques such as screen printing, flexography, and lithography, and even digitally. **Applications** include architectural graphics, point of purchase, advertising, labels, packaging, credit cards, and gift cards.

MC# 7099-01
Light Tape®
Electro-LuminX® Lighting Corporation
www.lighttape.com

Electroluminescent flexible film that is suitable for indoor and outdoor use. Its power consumption is a fraction of that found in LED lighting appliances, allowing it to be operated at significant distances from its power source. The material is manufactured by using a fluoropolymer barrier film to encapsulate phosphors, indium tin oxide proprietary chemistry, and precious metal electrodes. Different filters are mixed with the phosphors to achieve different colorations. **Applications** include architectural accents and backlighting, branding campaigns, billboards, event and theatrical lighting, safety/egress lighting, and automotive/aeronautical/marine accent and signaling applications.

MC# 7106-02
Thinfilm NFC Barcode
Thinfilm Electronics ASA
www.thinfilm.no

Printed NFC (near field communication) tags that are a combination of an antenna, a radio chip, and some data memory. The tag enables the short-range wireless transfer of data between two devices. This is the first NFC interface produced using a printed electronics process, so it is low-cost in comparison to NFCs produced by etching or bonding processes. In addition, the size of data memory is low (128 bits), instead relying on intelligence being stored in a "cloud," thus also keeping costs low. **Applications** include smart packaging for in-store or at-home promotions and loss-prevention tags.

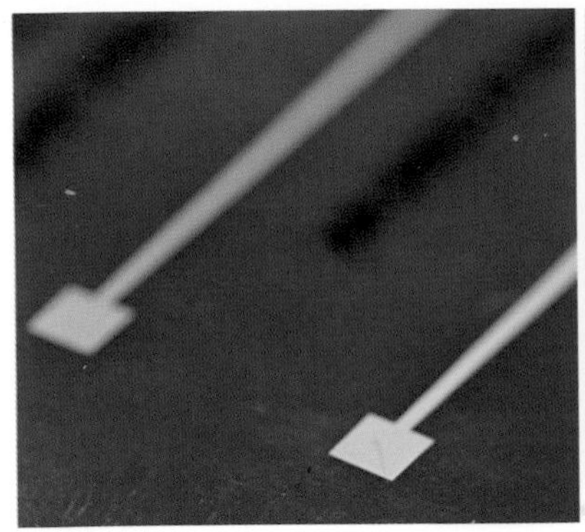

MC# 7167-01
Metalon® Inks
NovaCentrix
www.novacentrix.com

Water-based conductive inks for additive manufacturing of printed electronics. These inks contain silver or copper as the conductive element and are intended for application to lightweight, flexible, low-temperature substrates such as paper, PET, and other plastics. The inks can be applied by screen or inkjet printing; the material is then dried, cured, and converted to a metal thin-film by proprietary PulseForge® tools, a standard oven, or IR (infrared heating) equipment. **Applications** include printed electronics such as photovoltaic devices, RFID, touch-panel displays, medical test strips, smart cards and labels, circuit boards, and advanced or tamper-resistant packaging.

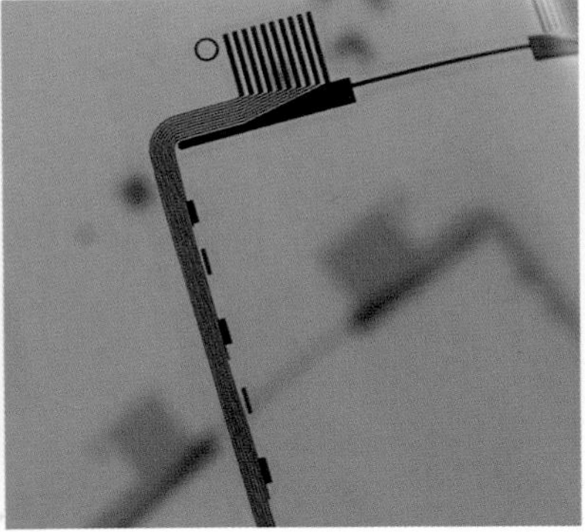

MC# 7207-01
CMB™
Canatu Ltd.
www.canatu.com

Flexible transparent films that conduct electricity for formable touch screens, displays, and touch-sensitive devices. These conductive films use ultra-fine deposited carbon nanotubes that are so small that they are effectively transparent, and are a combination of nanotube and fullerene, creating Carbon Nanobuds® that offer improved conductivity. They are deposited on to a clear polyester (PET) substrate using Direct Dry Printing® with no prior chemical cleaning required. The surface is claimed to be the most flexible transparent conductor on the market. **Applications** include mobile phones, tablets, e-readers, navigation devices, home appliances, and wearable electronics.

MC# 7277-01
Ynvisible Printed Electrochromic Display
YD Ynvisible, S.A.
www.ynvisible.com

Transparent, flexible printed display containing electrochromic ink that reveals its color when an electric charge is applied. In comparison to conventional non-light-emitting LCD-based displays or EPD (electronic paper displays, e-paper), this material has a lower power consumption and lower production costs. Screen and inkjet printing are used to create the displays; fine lines are possible, with the best results achieved in line widths (dot sizes) commonly used in the printing of graphics. **Applications** include interactive displays for promotional materials, greetings cards, magazine inserts and novelties, merchandise displays, POS signage, and consumer packaged goods.

MC# 7278-01
Active Paper™
The Active Paper Company Oy
www.theactivepaper.com

Proprietary paper and ink technology that rapidly reveals a hidden message when the paper becomes wet. In comparison to similar technologies, namely hydrochromic inks, which merely block the color underneath, this material is able to create a more uniform look and feel to the surface of the paper. By controlling the hydrophobic and hydrophilic properties of the paper through embedded channels, the flow of liquid is guided toward the areas of the paper where the hidden inks are printed, causing them to activate and reveal their color. **Applications** include packaging, gifts, and security labeling.

MC# 7279-01
Chameleon™
Rollprint Packaging Products Inc.
www.rollprint.com

A coextruded coated sealant film for heat-sealed plastic adhesion with color. This sealant layer increases the visual impact of the peel indicator by creating a color-changing effect. The top film of a given color, once opened, will leave behind a stripe of a different color on the bottom film where the two had been bonded. This aids in the identification of tampering or seal compromise and provides an additional anti-counterfeiting measure. **Applications** include heat-sealable films for packaging of medical equipment and food.

MC# 7285-01
Solidskin 3D Optic Film
3SMK Co., Ltd.
www.3smk.net

Flexible translucent polymer film that gives a 3D optical effect for security applications. Using a lenticular-type construction of the surface, the printed pattern on the sheet appears to move when it is tilted. Most existing standard lenticular sheets feature parallel raised ridges that move the position of the image printed on the underside. This version uses very fine-scale "micro lenses"—raised domes approximately 0.07 mm (0.03 in.) in diameter—that also create this effect, but on a higher resolution to a more dramatic degree. **Applications** are for anti-counterfeiting, as well as covers for personal electronics.

MC# 1930-08
Tyvek® for Graphical Applications
DuPont Nonwovens
www.graphics.dupont.com

A decorative non-woven textile that is lighter and stronger than paper and more versatile than fabric, specifically produced for graphical applications. Composed of high-density polyethylene (HDPE), this material is durable, dimensionally stable, and tear-resistant, and can withstand repeated folding and flexing. This material is manufactured in a flash spinning process and treated with Corona discharge (plasma treatment to alter the surface energy) and an antistatic treatment. The material can be color-coated (by converters—companies producing packaging from roll and sheet stock) in three standard colors: metallic silver, gold, and amber. **Applications** include packaging, stationery, and other decorative products.

MC# 2687-20
True Paper Croc
Maya Romanoff Corporation
www.mayaromanoff.com

This wall covering is composed of latex-saturated paper with an embossed "crocodile-skin" effect. All inks used in its manufacturing are chlorine-free, pH neutral, and completely non-toxic, with no ozone-depleting chemicals. The paper passes the international Consumer Product Safety Improvement Act (CPSIA) testing requirements to ensure phthalate- and lead-free materials. **Applications** include wall coverings, but with potential for use in packaging.

MC# 4805-06
TL644 Film
Bemis Associates Inc.
www.bemisworldwide.com

Soft, flexible two-layer elastomeric polyurethane film which is able to modify a fabric's "strength" in strategic locations to achieve shaping or increased compression. It has excellent wash resistance and a low activation temperature so it can be used with heat-sensitive fabrics. It adheres to a variety of fabrics using a heated press. Supplied on a matte release paper, it can be laser-cut to any pattern or shape and bonded to select panels or assemblies as required. Six colors are available, all in a matte finish: black, slate, smoke, white, sapphire, and copper. **Applications** include compression and performance wear.

MC# 4879-03
ENIGMA® Reflections featuring ClearTint®
Clariant Masterbatches Division
www.clariant.masterbatches.com

Additives to achieve sparkling, clear colors and reflectivity effects in polypropylene (PP) -based plastics. PP is a semi-crystalline material that tends to have a hazy or milky appearance in its natural state. It is notoriously difficult to color because many dyes tend to migrate out of the material and pigments usually reduce clarity. These colorants have been added to this clarified PP that are very stable and do not diminish the clarity of the transparent PP. A wide range of colors and effects is possible, including glittery. **Applications** include molded parts, packaging, consumer goods, and storage containers.

MC# 5429-09
Curious Metallics
Arjowiggins Creative Papers
www.curiousstory.com

100% recycled paper with metal-like glittering finish and excellent lightfast properties. It is suitable for laser- and ink-printing techniques (offset printing, screen printing, hot foil printing, embossing, UV finishing) as well as folding and creasing. It is available in twenty-three colors, has been certified by the FSC, and is acid-, lignin-, and elemental chlorine-free. It is coated on both sides with patented coating designed to achieve excellent printability. **Applications** include envelopes for high-quality brochures, stationery, menus, brochures, and luxury packaging.

MC# 7000-04
Terratek® WC
Green Dot Holdings LLC
www.greendotpure.com

A polymer composite that blends recycled polypropylene (PP) and fine pine-wood particles (wood flour) from lumber-manufacturing waste. The combination of the two materials creates a material that is compatible with standard plastic manufacturing, while minimizing polymer usage and providing a unique "speckled" aesthetic. The wood particulate serves as both filler and structural reinforcement for the polymer. Two grades are available, tuned for the processing requirements of injection molding and extrusion. **Applications** include decking, furniture, structural components, toys, housewares, and consumer goods.

MC# 6448-01
think4D®
think4D™
www.think-4d.com

Post-print process that uses patented technology and high-speed automated production technology to lend shape and texture to two-dimensional representations. This highly customizable process allows for full coverage custom textures and detailed relief to any depth plus flexible finishing capabilities such as hot and cold foil stamping, metallic inks and coatings, and spot gloss and matte coatings. **Applications** are for advertising, packaging, book covers, direct marketing, POP displays, brochures, and wallpaper.

MC# 7057-01
Moldable Felt (VNF5107)
Nordifa AB
www.nordifa.se

Lightweight, moldable polyester felt material that has vibration- and sound-absorbing properties. The soft, plush feel of the felt is maintained even after molding. The fibers are thoroughly mixed and needled into a felt that can be molded into any desired form. After molding, the product can be cut into any shape by a water jet or knife. **Applications** include furniture, sound absorption, orthopedic inlays, and automotive interiors.

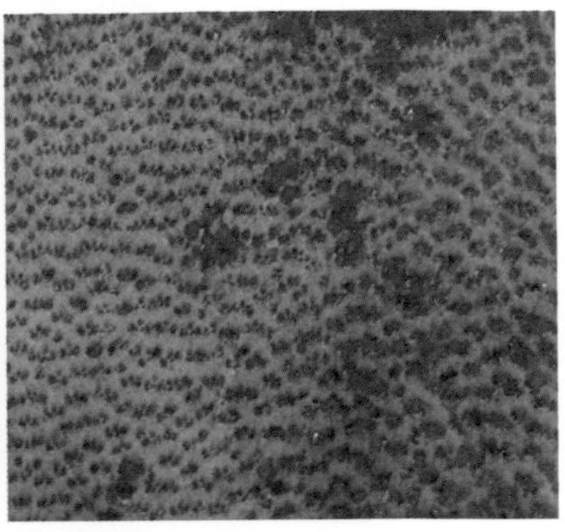

MC# 7108-01
Goatskin Parchment/Vellum
KARE LEATHER & PARCHMENT
www.parsomen.com

Thin parchment derived from goatskin that is handcrafted and does not go through tanning processes of any kind, making each piece unique. Animal skin is thoroughly flayed before being treated with different chemical additives to affect the material's texture and appearance. Customizations pertaining to the material's shape, size, color, and texture are all available. **Applications** include interior design and decoration such as wall and furniture coverings and lampshades.

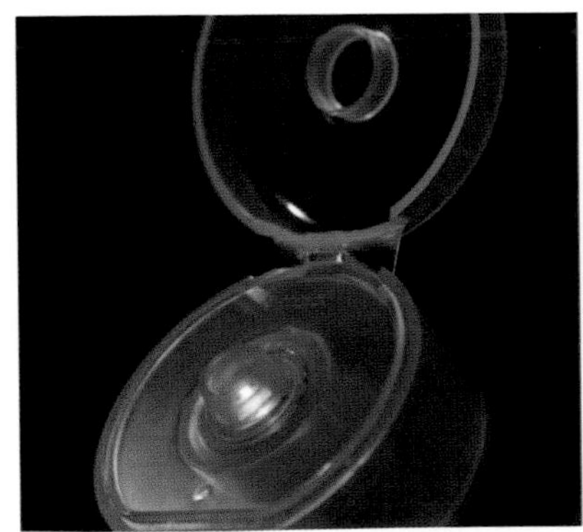

MC# 7110-01
Injection Molded Tube with In-Mold Label
Viva Healthcare Packaging (Canada) Ltd.
www.viva-healthcare.com

Injection-molded polypropylene (PP) tubes that are sheathed in a high-resolution printed PP label through the utilization of in-mold decoration techniques to give recyclable packaging. Continuous labeling creates a seamless image and allows for greater complexity of film visuals, especially through the incorporation of foils. Additionally the incorporation of the label within the molding process increases image durability against scuffs through the incorporation of an over-laminate. **Applications** include packaging for personal care, cosmetics, food, and industrial products.

MC# 7120-01
Oneplot Technologies & Photoinox
Oneplot S.r.l.
www.oneplot.com

Process for translating the colors and shadows of photographic images into an array of various sized holes through the use of a proprietary software program. The diameter of the holes as well as their density within the array will vary depending on the intended viewing distance and angle. The programmed pattern can be punched, laser-cut, or photochemically etched into a number of opaque and translucent materials including metal, paper, cardstock, glass, and polymer sheets. **Applications** include furniture inserts, wall panels, room dividers, lampshades, signage, retail displays, and stationery.

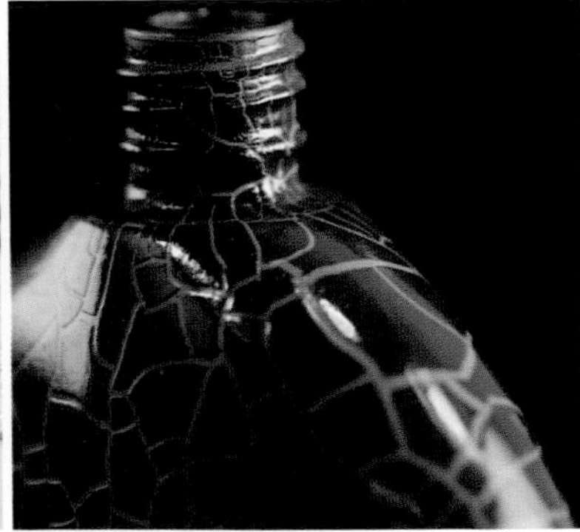

MC# 7246-01
Crackled Lacquering
SGD North America Inc.
www.sgdgroup.com

This lacquering process layers two different-colored lacquers together for a decorative cracked effect, the base layer becoming visible through the cracks of the top layer of color. Each application of this technique is unique, with cracks varying in size and distribution. This effect is achieved by carefully designed chemical incompatibility of the two layers, causing them to separate and crack as they set. The size of the cracks will vary due to substrate curvature. **Applications** include bottle decoration, packaging, furniture, and accessories.

MC# 7135-01
Decomet®
Schlenk Metallic Pigments GmbH
www.schlenk.com

Mirror-like VMPs (vacuum-metallized pigment) for coil coatings. Thanks to its flawless surface, realized in a vacuum environment, the pigment accomplishes highly reflective coatings. Depending on the applied type, it enables the user to create platinum, silver, chrome, or mirror-like surfaces. The pigments have an extremely smooth surface, leading to optimum reflection and extraordinary brilliance. The material is applied by spray-coating on the reverse side of transparent acrylic sheet and can be applied to all transparent surfaces. **Applications** include printing inks and coatings.

MC# 7171-01
Dejection Molding
Manuel Jouvin
www.manueljouvin.com

Dyed snail excrement (dejections) that has been altered by a diet of colored paper. The snails are fed a strict regime of colored cellulose paper, and because they cannot assimilate the paper's pigments their bodies reject the color of the ingested paper, creating colored excrement. Made from natural resources, this material is recyclable, biodegradable, and lightweight. It is available in many colors as different-colored papers with a unique texture and finish. Customization is available for paper color. **Applications** include floor tiles and packaging.

MC# 7192-01
Cocoform®
Enkev Group B.V.
www.enkev.com

Moldable fibrous composite that offers easy formability of complex shapes with good dimensional stability. The composite is a combination of 60% coir fibers and 40% natural latex that allows the fibers to be exposed and constitute the surface texture of the part. The parts are impact-resistant, have good elastic recovery, and are smooth to the touch. Both materials are rapidly renewable and the molded parts are biodegradable and compostable. **Applications** include packaging and interior parts for upholstery.

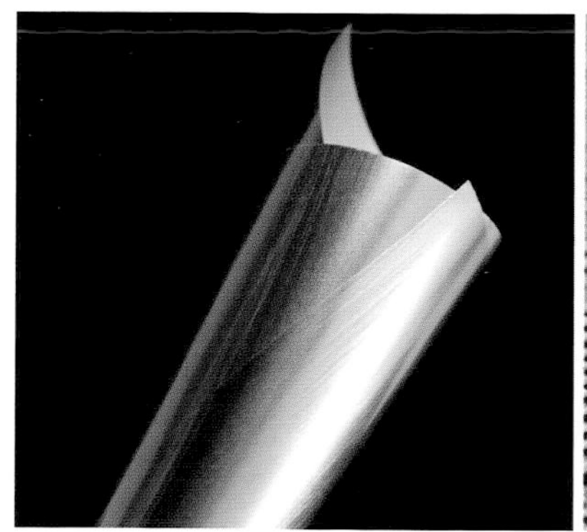

MC# 7214-01
Brushed Foil
Washin Chemical Industry Co., Ltd.
www.washin-chemical.co.jp

Thin metallized acrylic foil with fine line patterns that can be hot-stamped. This foil minimizes dust during the stamping process thanks to its fine edge definition and has high chemical and abrasion resistance as well as resistance to heat and cold. It is offered in silver, matte silver, gold, matte gold, chrome, matte chrome, black, and matte black in two brushed patterns with the option to customize color. **Applications** include decoration of plastic for packaging of consumer products, electronics, cosmetics, and decorative items.

MC# 7225-01
Scodix SENSE™
Scodix Ltd.
www.scodix.com

This printing process deposits a layer of clear polymer on to printed graphics to provide haptic effects to flat substrates. Raised elements can be up to 100 times the height of common selective varnishes, and are deposited with extremely high precision. To create perfect registration of the clear polymer over the printed graphics, the software utilizes the registration dots from the graphic printing to align the texture print. The clear polymer is applied via an inkjet press; the printer head deposits one billion droplets every four seconds. **Applications** include packaging, stationery, hang tags, business cards, magazines, and books.

MC# 7266-01
Dolce Foil
Katani Sangyo Co., Ltd.
www.katani.co.jp

Metallized polyester (PET) hot-stamping foil for surface decoration of plastics. The foils can be applied to molded or extruded surfaces using a hot-stamping process that transfers the foil using heat, pressure, and an adhesive. They are available in four colors: silver, green, blue, and magenta. An additional coating can be added to provide further protection, and the foils may also be printed on to. The color of the foil can be customized. **Applications** include cosmetics, labels, and consumer electronics.

RESOURCES

RESEARCH CENTERS & LABS

The Sonoco Institute of Packaging Design and Graphics Clemson University
311 Harris A. Smith Building
Clemson, SC 29634
United States
T: +1 864 656 4690
www.clemson.edu/sonoco_institute

Georgia Tech 3D Systems Packaging Research Center
813 Ferst Drive
Atlanta, GA 30332
United States
T: +1 404 894 9097
www.prc.gatech.edu

School of Packaging Michigan State University
130 Packaging Building
448 Wilson Road
East Lansing, MI 48824-1223
United States
T: +1 517 355 9580
www.packaging.msu.edu

Self-Assembly Lab SUDT-MIT International Design Center
Building N52, 3rd Floor
77 Massachusetts Avenue
Cambridge, MA 02139
United States
T: +1 617 253 3799
E: info@selfassemblylab.net
www.sutd.mit.edu

PROFESSIONAL ORGANIZATIONS

www.packworld.com/packaging-associations

ABRE Brazilian Packaging Association
Rua Oscar Freire 379
15° andar – cj. 152
Cerqueira César 01426-001
São Paulo – SP
Brazil
T: +55 11 3082 9722
abre@abre.org.br
www.abre.org.br

Alliance Graphique Internationale (AGI)
Frenzi Biedermann
AGI Secretariat
Pfingstweidstrasse 6
8005 Zurich
Switzerland
T: +41 71 393 58 48
E: info@a-g-i.org
www.a-g-i.org

American Institute of Graphic Arts (AIGA)
233 Broadway
17th Floor
New York, NY 10279
T: +1 212 807 1990
www.aiga.org

CENTREXPO S.p.A.
Centro Mostre Specializzate
Corso Sempione 4
20154 Milano
Italy
P.I. 04687000150
P: +39 02 319109.1
E: centrexpo@centrexpo.it
www.centrexpo.it

Institute of Packaging Professionals (IoPP)
1833 Centre Point Circle, Suite 123
Naperville, IL 60563
United States
P: +1 630 544 5050
E: info@iopp.org
www.iopp.org

International Association of Packaging Research Institutes (IAPRI)
P: +44 782436329
www.iapriweb.org

Ipack-Ima SpA
Corso Sempione 4
20154 Milan
Italy
P.I. 01620110153
P: +39 02 3191091
E: ipackima@ipackima.it
www.ipackima.it

LuxePack
Information
IDICE SAS
33 Cours de Verdun
01106 OYONNAX
France
T: +33 (0)4 74 73 42 33
E: info@idice.fr
www.luxepack.com

Organization
IDICE MC
13 Bvd Princesse Charlotte
Le Victoria, Bât D
98000 MONACO
T: +377 97 77 85 60
E: info@idice.mc
www.luxepack.com

Reusable Packaging Association
P.O. Box 369
Linden, VA 22642
T: +1 540 631 0773
E: info@reusables.org
www.reusables.org

World Packaging Organization
1833 Centre Point Circle, Suite 123
Naperville, IL 60563
United States
P: +1 630 596 9007
E: info@worldpackaging.org
www.worldpackaging.org

BLOGS AND WEBSITES

Ambalaj
www.ambalaj.se

BP&O
www.bpando.org

Brand New
www.underconsideration.com/brandnew/

createid
www.createid.com

Dexigner
www.dexigner.com

Lovely Package
www.lovelypackage.com

Packaging Design Archive
www.packagingdesignarchive.org

Packaging Design Served
www.packagingserved.com

Packaging of the World
www.packagingoftheworld.com

Pinterest
www.pinterest.com

The Dieline
www.thedieline.com

The Packaging Design Blog
www.thepackagingdesignblog.com

JOURNALS & MAGAZINES: PRINT & DIGITAL

www.worldpackaging.org

Beauty Packaging
www.beautypackaging.com

Brand Packaging
www.brandpackaging.com

Package Design
www.packagedesignmag.com

Packaging Digest
www.packagingdigest.com

Packaging Europe
www.packagingeurope.com

Packaging Today
www.globaltrademedia.com

Packaging World
www.packworld.com

TRADE SHOW & EVENTS

January
Packaging Design Matters
www.packagedesignmatters.com

February
The Packaging Conference
www.thepackagingconference.com

April
Luxe Pack Shanghai
www.luxepackshanghai.com

Packaging Innovations
www.easyfairs.com

The Reusable Packaging Forum
www.reusables.org

May
IPACK-IMA
www.ipack-ima.com

Luxe Pack New York
www.luxepacknewyork.com

National Association of Container Distributors Annual Convention
www.nacd.net

The Automation Conference
www.theautomationconference.com

The Dieline Conference
www.thedieline.com/events/

June
American Packaging Summit
www.packaging-event.com

Expo Pack Mexico
www.expopack.com.mx

FMI Connect and InterBev
www.fmiconnect.net
www.interbev.com

HBA Global Expo & Conference
www.hbaglobal.packagingdigest.com

Propak Asia
www.propakasia.com

Summer Fancy Food Show
www.specialtyfood.com

Taipei Pack
www.taipeipack.com.tw

July
ProPack China
www.propakchina.com

September
Innovations in Food Packaging
www.foodpackconference.com

Luxury Packaging
www.easyfairs.com

Pack Expo Las Vegas
www.packexpolasvegas.com

October
Luxe Pack Monaco
www.luxepack.com

Packaging that Sells
www.packagingthatsells.com

November
EMBALLAGE
www.all4pack.com/emballage

ADDITIONAL READING

Albers, Josef. *Interaction of Color: 50th Anniversary Edition*. Yale University Press, New Haven and London, 2013

Aldridge, Steven and Laurel Miller. *Why Shrinkwrap a Cucumber? The Complete Guide to Environmental Packaging*. Laurence King, London, 2012

Ambrose, Gavin. *Packaging the Brand: The Relationship Between Packaging Design and Brand Identity*. AVA Publishing, Worthing, 2011

Benyus, Janine M. *Biomimicry: Innovation Inspired by Nature*. Morrow, New York, 2009

Berger, Warren. *A More Beautiful Question*. Bloomsbury, London and New York, 2014

Beylerian, George M., Andrew H. Dent and Anita Moryadas. *Material ConneXion: The Global Resource of New and Innovative Materials for Architects, Artists and Designers*. Wiley, Chichester, 2005

Beylerian, George M., Andrew H. Dent and Bradley Quinn. *Ultra Materials: How Materials Innovation is Changing the World*. Thames & Hudson, London and New York, 2007

Cross, Nigel. *Design Thinking: Understanding How Designers Think and Work*. Berg, Oxford, 2011

DuPuis, Steven and John Silva. *Package Design Workbook: The Art and Science of Successful Packaging*. Rockport Publishers, 2011

Eckstut, Joann and Arielle Eckstut. *The Secret Language of Color: Science, Nature, History, Culture, Beauty of Red, Orange, Yellow, Green, Blue, & Violet*. Black Dog & Leventhal Publishers, New York, 2013

Emblem, Anne and Henry Emblem. *Packaging Technology: Fundamentals, Materials and Processes*. Woodhead Publishing, Cambridge, 2012

Esslinger, Hartmut. *Design Forward: Creative Strategies for Sustainable Change*. Arnoldsche Verlagsanstalt, Stuttgart, 2013

Fili, Louise. *Elegantissima: The Design and Typography of Louise Fili*. Princeton Architectural Press, New York, 2012

Fletcher, Alan. *The Art of Looking Sideways*. Phaidon Press, London, 2001

Gibbs, Andrew. *Box Bottle Bag: The World's Best Package Designs from TheDieline.com*. HOW Books, 2010

Hara, Kenya. *Designing Design*. Lars Muller Publishers, Switzerland, 2015

Hofmann, Armin. *Graphic Design Manual: Principles and Practice*. Braun Publishing, Switzerland, 2001

Jackson, Paul. *Cut and Fold Techniques for Pop-Up Designs*. Laurence King, London, 2014

Jackson, Paul. *Structural Packaging: Design Your Own Boxes and 3D Forms*. Laurence King, London, 2012

Kelley, Tom and David Kelley. *Creative Confidence: Unleashing the Creative Potential Within Us All*. Crown Business, New York, 2013

Klimchuk, Marianne R. and Sandra A. Krasovec. *Packaging Design: Successful Product Branding From Concept to Shelf*. Wiley, Chichester, 2013

Lovell, Sophie and Klaus Kemp. *Dieter Rams: As Little Design as Possible*. Phaidon Press, London, 2011

Lu, Daniel and C.P. Wong. *Materials for Advanced Packaging*. Springer, New York, 2009

Lupton, Ellen. *Thinking with Type: A Primer for Designers: A Critical Guide for Designers, Writers, Editors, & Students*. Princeton Architectural Press, New York, 2004

McDonough, William and Michael Braungart. *Cradle to Cradle: Remaking the Way We Make Things*. North Point Press, New York, 2002

McDonough, William and Michael Braungart. *The Upcycle: Beyond Sustainability–Designing for Abundance*. North Point Press, New York, 2013

Millman, Debbie. *Brand Thinking and Other Noble Pursuits*. Allworth Press, New York, 2013

Neumeier, Marty. *The Brand Gap: How to Bridge the Distance Between Business Strategy and Design*. New Riders, Berkeley, CA, 2005

Norman, Donald A. *The Design of Everyday Things*. Basic Books, New York, 2013

Oka, Hideyuki. *How to Wrap Five Eggs: Traditional Japanese Packaging*. Weatherhill, Boston, 2008

Papanek, Victor. *Design for the Real World: Human Ecology and Social Change*. Academy Chicago Publishers, Chicago, 2009

Pentawards and Julius Wiedemann. *The Package Design Book 3*. Taschen, Cologne, 2014

Rand, Paul. *Thoughts on Design*. Chronicle Books, San Francisco, 2014

Roth, Lászlo and George L. Wybenga. *The Packaging Designer's Book of Patterns*. Wiley, Chichester, 2012

Sudjic, Deyan. *The Language of Things: Understanding the World of Desirable Objects*. W. W. Norton, New York, 2009

Thompson, Rob. *Graphics and Packaging Production*. Thames & Hudson, London and New York, 2012

Twede, Diana and Susan E.M. Selke. *Cartons, Crates and Corrugated Board: Handbook of Paper and Wood Packaging Technology*. DEStech Publications Inc., Lancaster, 2004

Wheeler, Alina. *Designing Brand Identity*. Wiley, Chichester, 2012

GLOSSARY

ABS plastic (acrylonitrile butadiene styrene)
A common thermoplastic that is a stiffer, more impact-resistant alternative to polystyrene. It is the plastic used for many consumer and household goods, and is a predominant plastic for interior components of cars.

additive manufacturing
A collective term for manufacturing that builds parts layer by layer. This includes 3D printing, as well as 3D knitting and multilayer lamination.

chitosan
A naturally occurring polysaccharide, which is harvested by treating shrimp and other crustacean shells with the alkali sodium hydroxide. Polysaccharides are naturally occurring carbohydrate polymers, including cellulose and starch. *See also* xylan.

CNC (Computer Numeric Control)
The automation of machine tools that are operated by precisely programmed commands encoded on a computer, as opposed to controlled manually via hand wheels or levers. Most modern machining processes are now CNC.

electrochromic
Changing color when submitted to a burst of electrical charge. The electricity causes a usually reversible reaction in the chemistry of the material. The phenomenon occurs in transition metal oxides such as tungsten oxide (WO_3), which is used in electrochromic windows (smart glass). The polymer polyanaline also undergoes this transformation, from pale yellow to dark green/black.

electroluminescent
Emitting light in response to the passage of an electric current or to a strong electric field. It is the result of radiative recombination of electrons and holes in a material, usually a semiconductor. The excited electrons release their energy as photons (light). Both organic and inorganic electroluminescent materials exist and include zinc sulphide doped with copper (green light), silver (bright blue light) or manganese (orange-red); naturally blue diamond; and indium phosphide.

fullerene
The collective term for a molecule of carbon in the form of a hollow sphere, ellipsoid, tube, and many other shapes. Spherical fullerenes are called buckyballs, as they resemble Buckminster Fuller's geodesic (ball-like) domes. Cylindrical fullerenes are called carbon nanotubes or buckytubes. Fullerenes are similar in molecular structure to graphite, which is composed of stacked graphene sheets of linked hexagonal rings; but they may also contain pentagonal (or sometimes heptagonal) rings.

HDPE
See PE

hydrogel
A network of polymer chains that are hydrophilic, and able to absorb and contain a large proportion of water. They are physically and structurally similar to natural tissue. Hydrogels are also utilized as colloidal gels with water as the dispersion medium.

LDPE
See PE

hPDPS (printed dopant polysilicon)
A process for printing silicon chips to give them electronic properties. PDPS technology is based on a hybrid manufacturing process that leverages print methods on polysilcon surfaces. The technology enables high-performance transistors that support the high-frequency RF circuitry required for NFC communications.

PE (polyethylene)
The most widely used polymer, based on the simple hydrocarbon molecule C_2H_4. This lightweight, translucent thermoplastic is used in grocery bags and geomembranes. High-density formulations (HDPE) are used for milk containers, pipes, and toys, low-density versions (LDPE) in films and rigid containers. It can be foamed, and cross-linked to make it more durable, or blended to make rubbery elastomers.

PET, PETE (polyethylene terephthalate)
The chemical name for the polyester widely used as a fabric fiber and in transparent drinks bottles and some food packaging. Polyesters are a family of thermoplastics and thermosets that contain an ester in their polymer chain. They can be natural or synthetic materials; some are biodegradable.

PHA (polyhydroxyalkanoate)
A range of linear polyesters that are thermoplastic or elastomeric, and are produced in nature by bacterial fermentation of sugar or lipids. The bacteria produce the polymer within their bodies to store carbon and energy. More than 150 different monomers can be combined within this family to generate materials with extremely different properties. These plastics are biodegradable and are used in the production of bioplastics for packaging and consumer goods.

PLA (polylactic acid)
A polyester-like plastic produced from the polymerization of renewable resources such as corn starch, tapioca, or sugarcane. It is certified as industrially compostable and is used in packaging and food service products and as a fiber for textiles.

PP (polypropylene)
Polypropylene is a translucent, semi-crystalline thermoplastic and one of the most widely used polymers thanks to a combination of low cost, easy processing, non-toxic chemistry, and durable performance. It is used in a wide variety of applications including packaging and labeling, textiles, stationery, reusable containers, laboratory equipment, loudspeakers, automotive components, and banknotes.

polyolefin
The family of commodity plastics that includes polypropylene (yogurt containers, some chairs and most living hinges) and polyethylene (milk containers and most disposable grocery bags). These plastics can also be blended with rubbers to create TPO, a thermoplastic polyolefin elastomer, used by the car industry as an alternative to vinyl dashboards.

PVOH (polyvinyl alcohol)
A water-soluble synthetic polymer used in paper-making, textiles, and as water-soluble packaging.

sintering
The bonding of particulates into a solid object through the application of heat and/or pressure, without melting or liquefying the material.

selective laser sintering
An additive manufacturing process in which a laser heats powdered material, fusing it into a solid object. *See also* sintering.

static dissipative
A material or additive that dissipates static electricity through electrical conductivity.

susceptor
A material with the ability to absorb electromagnetic energy and convert it to heat (which is sometimes designed to be re-emitted as infrared thermal radiation). The input energy is typically radiofrequency or microwave radiation used in industrial heating processes, or sometimes in microwave cooking.

VOC (volatile organic compound)
These are organic compounds that evaporate in large amounts from a product owing to the compound's low melting point. These gases are not acutely toxic, but some can have harmful effects on humans and animals, and are regulated. Glues, binders, and paints can contain VOCs such as formaldehyde or solvents that have melting points below room temperature; VOCs are also emitted from plants such as wood. Most scents and odors we smell are VOCs.

xylan
Xylans are polysaccharides made from units of xylose (a type of monosaccharide). They are found in plant cell walls and some algae. Polysaccharides are naturally occurring carbohydrate polymers, including cellulose and starch. *See also* chitosan.

ABBREVIATIONS

CFIA Canadian Food Inspection Agency
FSC Forest Stewardship Council
HVAC Heating, ventilating, and air conditioning
IDEA Individuals with Disabilities Education Act
LEED Leadership in Energy and Environmental Design
NFC Near field communication
QR code Quick response code
RFID Radio frequency identification

CONTRIBUTOR CONTACTS

Airopack Technology Group
Van Schijndelstraat 7
5145 RE Waalwijk
The Netherlands
T: +31 416 30 08 00
E: info@airopack.com
www.airopackgroup.com

Shigeno Araki Design & Co.
2-11-26-9F, Minamisenba
Chuo-ku, Osaka, 542-0081
Japan
T: +81 6 6245 1800
E: design@nagasakido.co.jp

Atelier BangBang
3081 Ontario Est, Local 420
Montréal, Québec, H1W 1N7
Canada
T: +1 514 298 6420
E: info@atelierbangbang.ca
www.atelierbangbang.ca

Muli Bazak
Israel
T: +972 546399610
E: shmuelbazak@gmail.com

Bemis
One Neenah Center, 4th Floor
P.O. Box 669
Neenah, WI 54957-0669
United States
T: +1 920 527 5000
E: contactBemis@bemis.com
www.bemis.com

Bijl Architecture
Suite 7 -100 Penshurst Street,
Willoughby, New South Wales, 2068
Australia
T: +61 2 9958 7950
E: info@bijlarchitecture.com.au
www.bijlarchitecture.com.au

Carré Noir
24 rue Salomon de Rothschild
92150 Suresnes
France
T : +33 (0)1 57 32 87 00
E: carrenoir@carre-noir.fr
www.carrenoir.com

Chris Curro
United States
E: chris@curro.cc

dBOD
INIT Unit 5,80
Jacob Bontiusplaats 9
1018 LL Amsterdam
The Netherlands
T: +31 (0)20 521 65 00
www.dbod.nl

Design Directions Pvt. Ltd.
828 Shivajinagar
"Rajeev", Bhandarkar Road, Lane #13
Pune: 411 004
India
T: +91 98220 40752
E: satish@designdirections.net
www.designdirections.net

Design Futures
Cantor Building
153 Arundel Street
Sheffield S1 2NU
United Kingdom
T: +44 (0)114 225 6750
E: info@dfgroup.co.uk
www.designfuturesgroup.com

Disappearing Package
512 Grand Oak Dr.
Hillsborough, NC 27278
United States
T: +1 312 315 3220
E: aaron@alocato.com
www.alocato.com

Ecologic Brands Inc.
One Kaiser Plaza, Suite 1440
Oakland, CA 94612
United States
T: 877-ECO-LOGC
E: info@ecologicbrands.com
www.ecologicbrands.com

Eco-Products
4755 Walnut Street
Boulder, CO, 80301
United States
T: +1 303 449 1876
www.ecoproducts.com

Ecovative Design
70 Cohoes Avenue
Green Island, NY 12183
United States
T: +1 518 273 3753
E: info@ecovativedesign.com
www.ecovativedesign.com

Enkev Natural Fibres
De Toek 2
P.O. Box 3
1130 AA Volendam
The Netherlands
T: +31(0)299 364355
E: office@enkev.com
www.enkev.com

F/Nazca Saatchi & Saatchi
Avenue Lebanon, 253
Ibirapuera, São Paulo, SP
CEP 04501-000
Brazil
T: +55 11 3059 4800
www.fnazca.com.br

FoldedPak Inc.
5885 E Stapleton Dr, N C-305
Denver, CO, 80216
United States
T: +1 303 329 8477
E: efreas@expandos.com
www.expandos.com

Marieke Frensemeier
INM - Leibniz-Institut für Neue Materialien GmbH
Campus D2 2, 66123 Saarbrücken
Germany
T: +49 (0)681-9300-312
E: mareike.frensemeier@inm-gmbh.de
www.inm-gmbh.de

fuseproject
1401 16th Street
San Francisco, CA 94103
United States
T: +1 415 908 1492
E: info@fuseproject.com
www.fuseproject.com

Brian Gartside
briangartside@gmail.com
www.briangartsi.de

Geometry Global
Hanauer Landstraße 147,
60314 Frankfurt
Germany
T: +49 (0)69 427282 700
E: info-germany@geometry.com
www.geometry.com/de

Mireia Gordi i Vila
E: mireiagordi@gmail.com
T: 07500 368469
www.cargocollective.com/please_draw

GrinOn Industries
7649 Winton Dr.
Indianapolis, IN 46268
United States
T: +1 317 388 5104
E: kjohnson@grinonindustries.com
www.BottomsUpBeer.com

Group 4
147 Simsbury Road
Avon, CT 06001
United States
T: 860.678.1570
E: msolomson@groupfour.com
www.groupfour.com

Grower's Cup
Coffeebrewer Nordic A/S
Kasmosevej 3
DK-5500, Middelfart
Denmark
T: +45 63 400 124
E: contact@growerscup.com
www.growerscup.com

Happy Tear
Bellmansgatan 26
118 47 Stockholm
Sweden
T: +46 8 410 241 40
E: info@happytear.com
www.happytear.com

Herman Miller Inc.
855 East Main Ave. PO Box 302
Zeeland, MI 49464-0302
United States
T: +1 616 654 3000
E: hmcustomercare@hermanmiller.com
www.hermanmiller.com

Hornall Anderson
710 2nd Avenue, Suite 1300
Seattle, WA 98104
United States
T: +1 206 467 5800
E: us@hornallanderson.com
www.hornallanderson.com

31 St Petersburgh Place
London, W2 4LA
United Kingdom
T: +44 (0)20 7313 7820
E: uk@hornallanderson.com
www.hornallanderson.com

Hyperform
Marcelo Coelho Studio
288 Norfolk St, Suite 2
Cambridge, MA 02139
United States
T: +1 857 928 1874
E: email@cmarcelo.com
www.cmarcelo.com/hyperform/

Imperial Paints LLC
PO Box 489
Fairforest, SC 29336
United States
T: +1 864 595 3840
E: info@Imperialpaintsllc.com
www.imperialpaintsllc.com

Innventia AB
Drottning Kristinas väg 61
Stockholm
Sweden
T: +46 (0)8 676 70 00
E: info@innventia.com
www.innventia.com

KIAN Branding Agency
Ostapovsky proezd 3
Moscow
Russia
T: +7 495 926 09 86
E: info@kian.ru
www.kian.ru

Kurion
2020 Main St., Suite 300
Irvine, CA 92614
United States
T: +1 949 398 6350
www.kurion.com

Laser Food
Avenida de los deportes 8
Alzira, 46600 Valencia
Spain
T: +34 606801756
E: jsanfelix@laserfood.es
www.laserfood.es

LINHARDT GmbH & Co. KG
Dr. Winterling Straße 40
D-94234 Viechtach
Germany
T: +49 (0) 9942 951 – 0
E: info@linhardt.com
www.linhardt.com

LiquiGlide
75 Sidney Street
5th floor
Cambridge, MA 02139
United States
E: info@liquiglide.com
www.liquiglide.com

Lowe Brindfors
Birger Jarlsgatan 57C,
SE-113 83 Stockholm,
Sweden
T: +46 8 566 255 00
E: info@lowebrindfors.se
www.lowebrindfors.se

MAP
37–42 Charlotte Road
London EC2A 3PG
United Kingdom
T: +44 (0)20 3376 5931
E: mail@mapprojectoffice.com
www.mapprojectoffice.com

mcgarrybowen São Paulo (formerly AGE Isobar)
R. Wisard, 298
Vila Madalena
São Paulo, SP
Brazil
T: +55 11 2173 0333
www.mcgarrybowen.com.br

Mead Westvaco Corporation
501 South 5th Street
Richmond, VA 23219-0501
United States
T: +1 804 444 1000
www.mwv.com

Txaber Mentxaka
Spain
E: txaber@txaber.net
www.txaber.net

Motiff Creative Studio
Finland
E: mikael(ät)motiff.fi
E: arttu(ät)motiff.fi
E: nikolo(ät)motiff.fi
E: juho(ät)motiff.fi
www.motiff.fi

NINE
Kungsgatan 27
111 56 Stockholm
Sweden
T: +46 (0)8 58 00 99 99
E: hello@nine.se
www.nine.se

Pangea Organics
3195 Sterling Circle
Suite 200
Boulder, CO 80301
United States
T: +1 877 679 5854
E: support@pangeaorganics.com
www.pangeaorganics.com

pi global Europe
1 Colville Mews
Lonsdale Road
London, W11 2AR,
United Kingdom
T: + 44 (0)20 7908 0808
E: helloln@piglobal.com
www.piglobal.com

Poron XRD
Rogers Corporation
One Technology Drive
Rogers, CT 06263
United States
T: +1 800.935.2940 / 607.786.8112
E: solutions@rogerscorp.com
www.poronxrd.com

Prestige Products Packaging
London Innovation Centre (Egham)
Procter & Gamble Technical Centres Limited
Rusham Park
Whitehall Lane
Egham
Surrey TW20 9NW
United Kingdom
T: +44 (0)1784 474900
www.pg.com

Cédric Ragot Design Studio
2 rue Navoiseau
93100 Montreuil
France
T: +33 (0)1 48 58 89 47
E: contact@cedricragot.com
www.cedricragot.com

RKS Design
350 Conejo Ridge Avenue
Thousand Oaks, CA 91361
United States
T: +1 805 370 1200
www.rksdesign.com

Satumaa Family Business
Bolinders plan 2
112 24 Stockholm
Sweden
T: +46 8580 80 900
E: marko.edfelt@sfbagency.com
www.familybusiness.se

ScaldoPack
Boulevard des Canadiens
118 B-7711 Dottignies
Belgium
T: +32 56 23 00 00
E: info@scaldopack.be
www.scaldopack.be

Scholz & Friends Group
Hanseatic Trade Center
Am Sandtorkai 76
20457 Hamburg
Germany
T: +49 40 3 76 81-0
E: info@s-f.com
www.scholz-and-friends.de

Self-Assembly Lab
Self-Assembly Lab, MIT
International Design Center
www.selfassemblylab.net

Smart Packaging Systems
9330 United Drive, Suite 100
Austin, Texas 78758
United States
T: +1 512 997 8620
E: lfrego@smartpackagingsystems.com
www.smartpackagingsystems.com

Sparkle Design Agency
47 rue Jules Guesde
92300 Levallois-Perret
France
T: +33 (0)1 47 30 73 50
E: contact.fr@sparkle-design.com
www.opend-group.com

T+ink
244 west 54th street, 9th floor
New York, NY 10019
United States
T: +1 212 957 2700
www.t-ink.com

Technotraf Wood Packaging SL.
P.I. Fabra i Coats Num 2.
Sant Vicenç de Torelló
08571 Barcelona
Spain
T: +34 938 593 403
E: hello@technotraf.com
www.technotraf.com

Teenage Engineering
SE Studio
Katarina Bangata 71
11642 Stockholm
Sweden
T: +46 8 599 953 72
www.teenageengineering.com

Tetra Pak Global Resources
Avenue Général-Guisan 70
1009 Pully
Switzerland
T: +41 21 729 21 11
www.tetrapak.com

The Absolut Company / Pernod Ricard
Årstaängsvägen 19B
S-117 43 Stockholm
Sweden
T: +46 8 744 70 00
www.theabsolutcompany.com

The Edible Project
5555 de Gaspé, Suite 105
Montréal H2T 2A3
Québec
Canada
T: +1 514 229 7778
E: diane.l.bisson@gmail.com
www.edibleproject.com

Thin Film Electronics ASA
Henrik Ibsens gate 100,
0255 Oslo
Norway
T: +47 23 27 51 59
www.thinfilm.no

Tomorrow Machine
Sweden
T: +46 768994470
T: +46 707656024
E: info@tomorrow machine.se
www.tomorrowmachine.se/

Unipal International
PO Box 1277,
Boise, ID 83701-1277
United States
T: +1 877 839 3943
E: info@unipalinternational.com
www.unipal.com

Henry Wang
United States
E: henry.wang92@gmail.com

Webb deVlam
91–94 Lower Marsh
London, SE1 7AB
United Kingdom
T: +44 (0)20 7202 4720
E: london@webbdevlam.com
www.webbdevlam.com

Whipsaw
434 South First Street
San Jose, CA 95113
United States
T: +1 408 297 9771
E: info@whipsaw.com
www.whipsaw.com

WikiFoods Inc.
161 First Street, 3rd Floor
Cambridge, MA 02139
United States
T: +1 617 491 6600
E: hello@wikifoodsinc.com
www.quantumdesigns.com

Ynvisible
Rua Mouzinho de Albuquerque 7
2070-104 Cartaxo
Portugal
T: +351 24 310 3174
E: info@ynvisible.com
www.ynvisible.com

Ziba Design
1044 NW 9th Ave
Portland, OR 97209
United States
T: +1 503 223 9606
E: connect@ziba.com
www.ziba.com

Material ConneXion
1271 Avenue of the Americas,
17th Floor
New York, NY 10020
United States
www.materialconnexion.com

Material ConneXion Italia
Via Sarca 339/f
20126 Milano
Italy
http://it.materialconnexion.com/

Material ConneXion Tokyo
METLIFE Aoyama Building
2-11-16 Minami Aoyama
Minato-ku, Tokyo 107-0062
Japan
www.jp.materialconnexion.com/

Material ConneXion Daegu
701-824 3F, Daegu Gyeongbuk
Design Center
107-4 Shincheon 3-dong,
Dong-gu Daegu
Korea
www.kr.materialconnexion.com/

Material ConneXion Skövde
IDC West Sweden AB
Box 133
SE-541 23 Skovde
Sweden
www.se.materialconnexion.com/

Material ConneXion Bangkok
6th Floor, The Emporium Shopping Complex
622 Sukhumvit 24
Bangkok 10110
Thailand
www.th.materialconnexion.com/

Every material possesses its own properties and characteristics, just as every visual technique has its own expressiveness and every physical form contains its own possibilities and limitations. A packaging designer's success is dependent upon his or her ability to transmit this knowledge with skill and sensitivity into a multi-sensorial experience that is unique to that product or brand. This book's primary goal, therefore, is to reveal for the young practitioner the extraordinary range of advanced materials available today, as well as their role in the creative process. Materials, whether found in nature or fabricated in a laboratory, offer their own unique path toward inspiration, experimentation, and beauty.

Because thinking and making are not alternatives but reciprocal forces within the design process, we have separated each chapter by a visual narrative, a spread that graphically depicts the journey a designer takes from concept to finished result. In individualistic ways, these pages suggest how a handful of designers around the world engage with materials, all toward creating innovative, efficient, and sustainable packaging solutions.

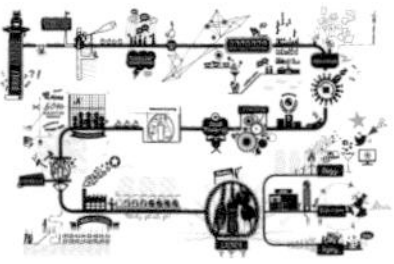

CARRÉ NOIR (PAGE 12)

Paris, France

No amount of ingenuity or creativity can create clear, distinctive, memorable design solutions from a confusing brief and a lack of strategic focus. This is why design is predominantly a way of organizing ideas, a process by which thought is made visible. Carré Noir's witty, revelatory process flow chart hints at the challenges and collaborative nature that inform the Paris-based firm's creative process. Their approach views a brand "like a person, it's the fruit of its parents' desire. One day it is born somewhere, it has its own character, it grows with age, and sometimes it gets married and has beautiful children. It grows older and can even disappear. Is a brand a person? Yes, but not like any other! A brand has the power to remain young. What if we had the power to make it immortal?"

FUSEPROJECT, YVES BÉHAR (PAGE 40)

San Francisco, USA

"In partnering with Puma, a leader in sports clothing, shoes, and products, we looked to create a game-changing packaging system that would greatly reduce environmental impact. The challenge was to look at one of the most difficult issues facing the retail industry in regard to sustainability: packaging, and specifically shoeboxes.

"Puma approached fuseproject to see what improvements could be made to reduce their footprint and reduce cost. A two-month exploration revealed where and how the current packaging was made, transported, packed, and stored through to delivery in the customer's hands. We documented and mapped the supply chain, indicating key requirements and areas for improvement at each step. With these insights, we designed the 'clever little bag,' a solution able to save vast amounts of of electricity, water, diesel fuel, and paper per year. The 'clever little bag' replaces the plastic shopping bag, and can also be repurposed for creative reuse. The bag is made of nonwoven polyester consisting of polypropylene, and eventually is also recyclable."

ZIBA (PAGE 66)

Portland, Oregon, USA

"RevMedx, a medical startup, developed a life-saving wound dressing: XStat. Ziba helped get this 'Fix-a-Flat® for wounds' into the hands of battlefield medics. Bleeding to death remains a leading risk for soldiers on the battlefield, and medics wanted a fast, foolproof hemorrhage solution that kept their hands free. Ziba designed the right delivery mechanism for XStat by modeling use cases with experienced medics and rapidly prototyping a wide variety of possible forms to suit their needs. We learned application had to be one-handed as well as ambidextrous. The final design delivers compressed sponges coated in an advanced hemostatic compound quickly and reliably without tools or setup. Life-or-death bleeding stops near-instantaneously, with no need for continued manual pressure. XStat combines aspects of traditional wound dressings, surgical interventions and drugs, leading the FDA to create an entirely new device category prior to its approval."

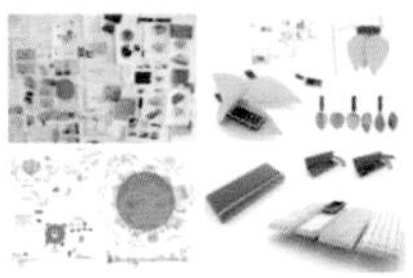

NINE (PAGE 92)

liselotte@nine.se

"Green Heart represents a brand-new sustainable packaging concept, material, and form factor for the mobile phone category. The material is called FiberForm® and was developed by NINE's partner Billerud. This was the first time FiberForm® had been used commercially, and the close collaboration between our packaging specialists, suppliers, and the client was a key factor in our success.

"NINE also developed the graphic identity and communication around the Green Heart project as well as the display concept for the launch event and retail presence, all in close collaboration with the design department at SonyEricsson. During the project NINE's Business Strategy Service crunched the numbers, and delivered a project white paper. The underlying insight and analysis was delivered by NINE's Future Research Service, with input from our global network of trend scouts and ethnographers.

"Future Research Service explores human behavior, attitudes, and values with the use of qualitative and ethnographic methods. By identifying and understanding un-met needs and desires we can help our clients develop future products and services that answer people's problems."

TEENAGE ENGINEERING (PAGE 126)

Stockholm, Sweden

"Teenage Engineering's products include the OP-1 portable synthesizer, Oplab musical experimental board and OD-11 cloud speaker. This spread is a mood board about our packaging system, with collages of materials and highlights of selected packaging. Our belief in high-quality products with timeless design, simplicity, and future-proof technology is reflected in everything related to our brand, including the packaging. We developed a modular system for our packaging and stationery—it's precise, technical, smart, beautiful, and is flexible enough to be applied widely. At Teenage Engineering, precision and spontaneity are equally important—that's why our design can be both classic and playful.

"Besides the basic functions of the packaging, we sometimes add extra features so that the packaging becomes a part of the product that you want to save. In this way its lifespan is extended. Packaging should be easy and friendly to use, and also present the aesthetic and quality of the product that's inside. We have pushed attention to detail to its furthest, and we always choose the best material and process when designing our packaging, to give the customer the best possible experience."

SHIGENO ARAKI (PAGE 150)

Osaka, Japan

"Visual communication is a process of making—of transforming ideas into tangible expressions. Elegant, minimal, modern yet traditional, the brand created by sweets manufacturer Zen Kashoin brings the production of traditional Japanese confectionery goods back to its roots, a philosophy carried through in its packaging. The company produces Japanese sweets based on the concept of 'returning flowers and confectionery to their authentic state.' We consider the meaning of 'authentic' as 'bare form.' By focusing on bringing out the material, texture, and ingredients, we made the packaging represent an eggshell, which softly and gently surrounds the product. And in consisting of a form that has been reduced to its essentials, the packaging is also thus an embodiment of the spirit of Zen, as it is manifest in the brand name. Bearing this in mind, we believe packaging indeed appeals to the senses."

DESIGN DIRECTIONS (PAGE 172)

Pune, India

A willingness to explore new ideas and new territory informs the work of Design Directions, and is exemplified by their reusable spool packaging: "First of all we had to understand the main reason why the spools were not reaching the customer in the desired orientation. The wire often became entangled while unwinding on a doffer. We had a meeting with our client and redefined the design brief. We realized that the main problem was the design of the existing packaging. A drum is always kept on its horizontal surface to avoid rolling. This meant that the spool was horizontal in such a position and the wire got tangled in transport due to the vibrations. We eventually redesigned the business model by designing a reusable packaging instead of a single-use one. This has a low impact on the environment. The rejection rate reduced drastically, and was almost nullified. The design helped to create a strong visual brand and commanded respect from the transporters."

PICTURE CREDITS

1: credits for the images on this page are listed under the corresponding projects; 2–3: Nikolo Kerimov; 6–8: fuseproject; 10: Design Futures; 11: fuseproject; 12–13: Carré Noir; 14: Enkev; 16–17: Jun Mitani; 18–19: Pangea Organics; 20–21: Bijl Architecture and Michael Ford; 22–23: Herman Miller Inc.; 24–25: György Kőrössy; 26–27: Group 4; 28–29: Courtesy of Laser Food; 30–33: Bob Beck; 34–35: Aaron Mickelson; 36–37: Simon Laliberté; 38–39: Ralf Schröder, Jinhi Kim, Scholz & Friends, Berlin GmbH; 40–41: fuseproject; 42: CP+B & Always Good Company; 44: Fidlock; 45: Industrial origami; 46–47: Linhardt GmbH & Co. KG; 48 49: Grower's Cup; 50–51: Peter Ascoli; 52–53: Ecologic Brands Inc.; 54–55: Smart Packaging Systems; 56–59: Courtesy of Pi Global; 60–61: Courtesy of Tetra Pak Packaging Solutions S.p.A.; 62–63: Lakshmi Card Clothing Manufacturing Company Pvt Ltd and Design Directions Pvt Ltd.; 64–65: Mark Serr; 66–67: Stuart Mullenberg Photography, LLC and Ziba Design Inc.; 68: Natural Machines; 70: 3D Systems Inc.; 71: Amazon.com Inc. 72–73: LiquiGlide; 74–75: Stuart Mullenberg Photography, LLC; 76–77: Eco-Products Inc.; 78–81: Kyle McMorrow; 82–83: Imperial Paints LLC; 84–85: Courtesy of Airopack; 86–87: Muli Bazak; 88–89; Kyle Johnson and GrinOn Industries; 90–91: Nikolo Kerimov; 92–93: Liselotte Tingvall of Nine; 94: Rogers Corporation; 96 left: Compadre; 96 right: Mitsubishi Chemical Corporation; 97: ScaldoPack; 98–101: Camilla Wirseen/Peepoople, Niklas Palmklint/Peepoople; 102–103: RKS Design; 104–107: Fedrigoni Deutschland GmbH; 108–109: Ecovative Design; 110–11: Kurion Inc.; 112–13: Unipal International; 114–15: MeadWestvaco Corporation; 116–17: Mareike Frensemeier, INM-Leibniz Institute; 118–19: Veuve Clicquot Ponsardin-LVMH; 120–21: Thin Film Electronics ASA; 122–23: FoldedPak Inc.; 124–25: Mireia Gordi i Vila; 126–27: Teenage Engineering; 128–29: YD Ynvisible, S.A.;130–31: MIT Self-Assembly Laboratory; 132–33: P&G Prestige design team; 134–35: Hornall Anderson UK; 136–37: T+Ink; 138–39: Fotograf Johan Olsson; 140–41: YD Ynvisible, S.A.; 142–45: KIAN; 146–47: Karl Knauer KG and Webb deVlam; 148–49: dBOD and iRIS WorldWide; 150–51: Shigeno Araki Design & Co.; 152: Marcelo Coelho and Skylar Tibbits; 154: Courtesy of Adidas; 155: Technotraf Wood Packaging SL.; 156–59: WikiFoods Inc.; 160–63: Absolut; 164–65: F/Nazca Saatchi & Saatchi; 166–67: Hornall Anderson UK; 168–69: mcgarrybowen São Paulo and Camp Nectar; 170–71: Txaber; 172–73: Lakshmi Card Clothing Manufacturing Company Pvt Ltd and Design Directions Pvt Ltd.; 174: Dan Bigelow Photography; 176–79: Dan Bigelow Photography; 180–99: all images in the Material Directory courtesy Material ConneXion

THE AUTHORS

Dr. Andrew Dent
Dr. Andrew Dent, Vice President, Library & Materials Research, plays a key role in the expansion of Material ConneXion's technical knowledge base. His research directs the implementation of consulting projects and the selection of innovative, sustainable, and advanced materials to Material ConneXion's Library, which currently has over 7,500 materials. From Whirlpool and Adidas to BMW, Proctor & Gamble and Ikea, Dr. Dent has helped numerous Fortune 500 companies develop or improve their products through the use of innovative materials. He is a frequent speaker on sustainable and innovative material strategies and is the co-author of the Material Innovation books series, which, to date, includes *Material Innovation: Architecture* and *Product Design*. He has also contributed to many magazines including *Business Week, Fast Company*, and the *Financial Times*. He received his Ph.D. in Materials Science from Cambridge University in the United Kingdom.

Leslie Sherr
Co-author with Dr. Andrew Dent of three books on material innovation, Leslie Sherr is a writer, editor, and brand strategist who champions the wonders of our built and designed world. As a brand strategist, she is the former Director of Marketing and Business Development at C&G Partners, where, working closely with founding partner Steff Geissbuhler, she helped shape the identities of universities, libraries, cultural institutions, and community development organizations. She has extensive experience as a marketing consultant with leading architecture and design firms, including Assouline, Chandelier, Munder-Skiles, The 7th Art and YARD, and has held positions at Carbone Smolan Agency and Desgrippes Gobé. She has a Bachelor of Fine Arts from SUNY Purchase and a Master of Science in Landscape Design from Columbia University.

Material ConneXion (materialconnexion.com) is a global materials and innovation consultancy that helps clients create the products and services of tomorrow through smart materials and design thinking. Material ConneXion, a SANDOW company, is the trusted adviser to Fortune 500 companies, as well as to any forward-thinking agencies and government entities seeking a creative, competitive, or sustainable edge. With seven locations—in Bangkok, Bilbao, Daegu, Milan, New York, Skövde, and Tokyo—Material ConneXion's international network of specialists provides a global, cross-industry perspective on materials, design, new product development, sustainability, and innovation. Material ConneXion maintains the world's largest subscription-based materials library, with more than 7,500 innovative materials and processes—an indispensable asset to a wide audience of users. The consulting division, ThinkLAB, works with clients to strategically incorporate trends, service, and innovation into their business models and products, while sister company Culture + Commerce represents the world's leading designers, including Philippe Starck and Marcel Wanders, in licensing their groundbreaking new products and projects.

ACKNOWLEDGMENTS

Although we appear as the authors of this book, it is in fact the work of many hands. This project has been in every sense a team effort, which would not have been possible without the vision of George M. Beylerian, founder of Material ConneXion, as well as the knowledge, skills, hard work, and dedication of many colleagues.

These pages are the culmination of many people's work and we sincerely thank those who were most closely connected with this book's creation. We are grateful to Michele Caniato, president, for bringing Material ConneXion and Thames & Hudson together to realize the full potential of this ambitious series, and to Adam I. Sandow, founder, chairman and chief executive officer of SANDOW Media Corporation, for his support. Our sincere thanks extend to Gabriella Vivaldi, Amanda Muchnick, Rachel Lexier-Nagle, and the Material ConneXion marketing department, especially Daniel Swartz, for working tirelessly to help bring this book to life. For his guidance: Matthew Kalishman; for contributing to the editorial team; Maider Irastorza, Fiona Anastas, Elizabeth Peterson, Sarah Hoit, Anuja Bagul, Anuja Joshi, Max Dreyer, Yaodi Wang, Jennifer Coppola, and Alejandra Kluger of Material ConneXion's Library team and ThinkLAB consulting division. Thank you also to the sales team at Material ConneXion, who have given their time and resources, and whose dedication is reflected in these pages.

We owe a special debt to all the outstanding designers and creative firms whose visions have so enriched our understanding of what is materially possible in the ever-thrilling field of packaging, and who have given so generously of their time, knowledge, and insight, especially Yves Béhar of fuseproject for his preface and John Kirkby of Design Futures Packaging at Sheffield Hallam University for his introduction to this volume. Our endless thanks also go out to the many contributors—photographers, media contacts, scientists, and manufacturers—whose creativity, research, and knowledge underpin every chapter in the book. It has not been possible here, for reasons of space, to include every single individual by name, but that does not lessen our gratitude to them.

Material ConneXion X10 has been created in collaboration with Thames & Hudson. Without Jamie Camplin's championing, this series would not have happened. We would like to extend our warmest appreciation to him, as well as to Ilona de Nemethy Sanigar, who offered invaluable direction and is a genuine pleasure to work with. We are also grateful to Johanna Neurath, design director, and Samuel Clark, senior designer, for the publication design, to Kirsty Seymour-Ure for her sensitive copyediting, and to Paul Hammond for his production control.